OCEAN FRONTIERS

Explorations by Oceanographers on Five Continents

OCEAN FRONTIERS

Explorations by Oceanographers on Five Continents

Edited by Elisabeth Mann Borgese

Harry N. Abrams, Inc., Publishers, New York

*Knowledge, mastered and shared,
could change the world.*

Comenius (1592–1670)

*This book is dedicated to the memory
of Roger Revelle, Oceanographer and
Humanist*

Editor: Adele Westbrook
Designers: Darilyn Lowe Carnes, Liz Trovato
Photo Editor: John K. Crowley

Library of Congress Cataloging-in-Publication Data
Ocean frontiers: explorations by oceanographers on
5 continents / edited by Elisabeth Mann Borgese.
p. cm.
Includes bibliographical references and index.
ISBN 0–8109–3665–8 (cloth)
1. Oceanography—Research   I. Borgese, Elisabeth Mann.
GC57.023   1992
551.46—dc20   91–32711
                                                    CIP

Title Page Green Sea Turtle in the waters off the
Galapagos Islands. Photo © 1990 Carl
Roessler

Contents

Preface

This book highlights many important developments in what is still a far from completely documented realm: human attempts to penetrate the secrets of the seas and oceans—the last and vast frontier on our planet.

By assembling this introduction to the history of a number of ocean research initiatives in twelve institutions on five continents, Professor Elisabeth Mann Borgese has again succeeded—with the help of a group of eminent world ocean experts—in drawing public attention to the importance of oceanography and marine scientific research. The examples she has selected combine to provide a fascinating account of the development of the world's marine community, reflecting the desire of nations and cultures throughout history to explore and exploit the vastness of the sea.

Oceanography, or marine science, is closely concerned with both the classic and modern uses of the sea and the associated technology. It contributes not only to the efficiency of fishing and to the safety of shipping but also to harbour construction and to protective measures for the safeguarding of populations living in lowlands, along river estuaries, and on the coasts of continents. Marine science has progressed greatly through the incorporation of many new technologies that have made it possible to exploit offshore oil and gas, to mine the minerals of the seabed and the ocean floor, and to explore the potential for marine waste disposal. By enhancing aquaculture invest-

ments and developing marine biotechnological studies, it serves to ensure additional resources for mankind.

Currently, the marine scientific community is very much involved in environmental science. In particular, it is playing a leading role in a series of global experiments designed to construct and verify numerical models of the earth's climate system that encompass not only the atmosphere but also the ocean and its physical and geobiochemical processes.

The potential of the ocean for mitigating the physical and chemical consequences of the increasing amounts of man-made greenhouse gases in the atmosphere is at present an unknown quantity since this part of our global environment is very inadequately observed and monitored. Special efforts in oceanography and related ocean services must therefore be directed towards this major area of uncertainty in our understanding of climate change.

For this reason, as well as for its role in developing the use of the coastal and high seas for the benefit of growing populations, oceanographic science must be promoted as one of the keys to the future well-being and welfare of humanity. If this book helps to sensitize governments and the general public to this need, it will have served a vital purpose over and above the valuable information and instruction I believe it will offer to the reader.

Federico Mayor
Director General, UNESCO

Coral reefs are of vital
importance to the
stabilization of coastlines and
as fish habitats. Photo:
courtesy Jun Oui

Kicker Rock in the Galapagos
Islands. Photo: © 1991 Carl
Roessler

Introduction

Elisabeth Mann Borgese

"O ceanography," the science or art of describing (*graphein*) the oceans, is the natural complement of "geography," the science or art of describing the earth (*gaea*), but until quite recently oceanography has lagged a few centuries behind its sister science. Our ocean maps were primitive, inexact, and full of lacunae.

Today oceanography has come of age. In a way it is now the cutting edge of the sciences describing our planet: It has transformed our traditional concepts of this planet and its evolution.

The ancients knew little about the seas and oceans, and what they knew was inextricably mixed with myth and folklore. Aristotle was an early expert in marine biology and in the chemistry of seawater, but he knew nothing about the basins containing these things. During the ages of exploration, in the fifteenth and sixteenth centuries, the oceans, or at least their surface, began to take on more realistic contours. Henry the Navigator, Infante of Portugal (1394–1460) founded the first interdisciplinary, international oceanographic institution near Sagres where he gathered mathematicians, astronomers, chart-makers, captains and chroniclers from Portugal and Spain, Venice, Genova, and the Arab countries. They worked on chart design; they improved compasses, astrolabes and quadrants; they compiled astronomical tables. They laid the foundation for the great voyages of exploration. Christopher Columbus was a student of Sagres. Sagres gave the world whole generations of great seafarers, from Gil Eanes, who rounded Cape Bohador in 1433, to Vasco da Gama, who landed in Calicut on the west coast of India, in 1498.

But the real beginning of modern oceanography is of very recent date, a mere 120 years ago, when, in 1872–76, H.M.S. *The Challenger* cruised the world ocean. It took the scientists of *The Challenger*—or the "philosophers," as they were called by the navy crew—ten years after the end of their voyage to complete the fifty thick volumes presenting the results of this first major oceanographic cruise. The report discussed with full detail of text and illustrations the currents, temperatures, depths and constituents of the oceans, the topography of the sea bottom, the geology and biology of its covering, and the animal life of the abyssal waters. That they could do this with the rudimentary technology at their disposal at the time, seems almost miraculous. That which is achieved now with power winches, piston corers, and remote controlled "boomerangs," they accomplished with hand-driven winches, lead-lines, and baskets. In the mid-Pacific, they lifted manganese nodules from a depth of five thousand meters, analyzed their composition, configurated their density and distribution, and made wild and wonderful guesses about their origin. Their computers they carried in their heads. Their navigational aids consisted of a compass, a chronometer (clock), and a quadrant; the pictures of the ocean's flora and fauna, produced today by deep-towed television and photo cameras, they etched with pencil and watercolor. One wonders whether human ingenuity is at its best producing high technology or obtaining the most complex and imposing results with the simplest means.

Since that time, oceanography has grown, expanded, and diversified at an astonishing rate. With the development of acoustic, seismic, and optical sensors, magnetology, radioactive fossil dating, remote sensing, deep-sea drilling technology, the development of deep-sea submersibles—as well as satellite technology, micro-electronics and computerized data processing—the human mind has been able to penetrate the ocean depths, right down to the ocean floor and its subsoil, through strata upon strata of sediments and basaltic rock, to the molten core of the planet earth. Thus it has been possible to reconstruct the evolution and transformation of ocean basins, to document with empirical data the theory of tectonic plates and continental drift, thus revealing that creation has not been a once-and-for-all happening but that it is an ongoing process regenerating and recycling the earth and its resources over eons.

Oceanography today is an interdisciplinary science, the breadth and depth of which is presented in Chapter 1. It comprises many subsectoral disciplines, including the study of ocean currents and submarine waves, and of marine/atmospheric interaction and meteorology, weather forecasting, and the safety of navigation. Another branch is marine chemistry and biochemistry, including studies that may have a crucial bearing on our understanding of "global change": on the one hand, the contribution of seafloor volcanism to the atmospheric carbon dioxide load (about thirty percent of all carbon dioxide in the atmosphere today

is generated by seafloor volcanic activity!); on the other hand, the capacity of the oceans to absorb carbon dioxide. Then there is marine biology and ecology, basic for the exploration and exploitation of the oceans' living resources, as well as the development of aquaculture, which today is penetrating and transforming the fishing industry at an ever accelerating pace.

The direct application of the marine sciences to the development of marine technology and the intensification and diversification of ocean industries is a dramatic spectacle. We may call it the penetration of the industrial revolution into the oceans—with vast implications for the global economy, the environment, military strategy, national and international organization. It is in this wider context that oceanography has assumed the fundamental importance it has today.

Marine industrial technology affects all uses of the sea.

The fishing industry has been transformed by space technology, satellite-generated data management, electronic charting, echo sounders and sonars, hydro-acoustic equipment for fish finding; fish-aggregating devices for concentrating the "harvest"; new materials and computerized and automated gear selection and application; mechanized processing at sea in huge factory ships, with bulk storage in chilled water tanks, and freezing and packaging on board. However, the impact of industrialized fishing on the traditional fisheries in coastal waters is often devastating, and more research is needed to safeguard these important food sources from over-exploitation, pollution, and coastal engineering.

As one commercially fished species after another succumbs to the onslaught of industrialized hunting, aquaculture—the husbanding of aquatic animals and the cultivating of aquatic plants—is beginning to overtake hunting and gathering from the wild. Aquaculture is carried out in sea water, fresh water, and brackish water, in ponds, rivers, lakes, reservoirs, canals, raceways, lagunae, enclosed seas, or even in the open oceans, as well as in totally controlled, artificial environments, like greenhouses or tanks. Aquaculture today generates fifteen percent of global fish and seaweed production, and this percentage is rapidly increasing, at about six percent per annum, with the potential of doubling in less than twelve years, before the end of the century. Aquaculture engineering—aquaculture technology, from pond construction and control of the environment to artificial spawning, the rearing of larvae, and economical feed production—is a new branch of high technology, affected increasingly by the dramatic developments in genetic engineering and bio-industrial processes. The potential is staggering, not only for increased food production, especially in regions where starvation is endemic, but also for pollution control, the pharmaceutical industry, and other industrial processes.

Environmental deterioration is an important factor inhibiting the full realization of the aquaculture potential. As explained in Chapter 9, oceanography is contributing to the alleviation of this problem. The ecosystems of aquaculture facilities, the composition of microbial populations in the pond (ranging from bacteria to protozoa), and the succession of populations accompanying the changing environmental conditions, have been studied in various types of aquaculture facilities. Some species of bacteria and protozoa were found to have the ability of maintaining or actually restoring the desirable water quality of ponds.

Many marine microorganisms produce biologically active substances: enzymes, enzyme inhibitors, and compounds having antibiotic, antitumor, antileukemic, and other pharmacological qualities. The isolation of strains of bacteria capable of determined biological activities, and genetic engineering to enhance these capabilities, is going to be of great importance, not only for the pharmaceutical industry but for a number of other industries as well: replacing chemical and mechanical with biological processes, e.g., for the cleanup of oil spills or the extraction of metals from ores through bacterial systems. The Mediterranean Blue Plan, compiled by the United Nations Environment Program over the past ten years, envisages a future bio-steel industry that could be developed, based on bacteria and solar energy.

The shipping industry has been revolutionized by new materials, computerized ship design, automated construction, satellite-linked navigational aids and containerization and unitization, leading to multimodal, global, door-to-door services. This has transformed not only the shipping industry as such, but also the development and management, even the function and role, of ports and harbors which have become nodal points in a global transport system, employing the most sophisticated computerized loading and off-loading technologies. Floating jetties, constructed of new materials, is another area of research and development in harbor construction. What oceanography is doing to make the construction of ports and harbors compatible with the conservation of the marine environment is indicated in Chapter 12.

New types of high-speed hovercraft, such as France's ADOC-12, riding on air cushions, are in the making. Industrial submersibles and underwater multi-service robots, such as France's SAGA 1 and ELIT, will intensify and accelerate deep-sea exploration and exploitation. The offshore oil industry has developed exploration systems involving seismic, acoustic, and optical instrumentation combined with data computerization that has increased its precision by a factor of several orders of magnitude, reducing the need for experimental drilling. Passing through the design of a series of exotic platforms of increasing sophistication, this development appears to be heading in the direction of sub-sea completion systems, making it possible to explore and exploit hydrocarbons at almost any depth and in almost any climate.

The development of offshore oil exploration and production technology is having spin-off effects on the development of deep-sea mining technology, which is still in an experimental stage and dependent on research and development in new materials, lasers, robotics, microelectronics, information technology and data handling, seismic, acoustic and optical technology, satellite-borne navigational aids and, probably, even bio-industrial processes for the processing of the minerals.

Deep-sea exploration technologies are, typically, "dual-purpose technologies." They enhance the peaceful uses of the oceans, the discovery of sea-mounts and submarine geysers and volcanoes, of chimneys and "thermal vents," and the strange colonies of unearthly living beings inhabiting these regions. They are also of fundamental importance in the military dimension such as submarine and anti-submarine warfare. The technology that precisely locates potato-sized nodules on the bottom of the sea at a depth of 5,000 meters, will also locate and identify hostile submarines. The study of currents, temperatures, densities, magnetism, and the submarine propagation of acoustic waves is basic to marine activities in peace as well as in war. The linkage between oceanography and military research is indicated in Chapter 11. The oceans have become the repository of about thirty percent of the world's nuclear arsenal with first- and second-strike capacity: a horrendous threat to peace, to economic development, and to the integrity of the marine environment.

It is the direct linkage to economic as well as strategic interests that gives oceanography today such enormous importance—it also generates two sets of major problems.

Oceanographic research may become the subject of fear and suspicion. The "philosophers" of H.M.S. *The Challenger* enjoyed the freedom of the seas and landed as welcome guests in any country. Today's researchers might be military spies or agents for industries bent on the exploitation of offshore resources unknown to the coastal State with jurisdiction over them. The new Law of the Seas subjects vast areas of ocean space to the jurisdiction of coastal States: an Exclusive Economic Zone of 200 miles; a continental shelf extending, in some cases, beyond 350 nautical miles from shore. Marine scientific research in these areas is no longer "free." Researchers need the consensus of the coastal State, which may entail bureaucratic delays as well as, perhaps, a shift of the focus of research: Oceanographic research should be relevant, henceforth, not only to the researching State and its institutions, but also to the coastal States under whose jurisdiction it is conducted, and which now have the right to participate in the research and share its results.

The Convention's "consent regime," however, is already being overtaken by the giddy pace of technological development. Remote sensing from satellites in outer space is becoming so accurate that it can explore, directly or indirectly, almost anything that used to be explored by ships *in situ*. And it can explore far wider areas in a far shorter time. Research from satellites is not subject to the regulations of the Law of the Sea Convention. The research thus carried out, once again, is "free." But it is reserved for the few and the rich.

The second set of problems arises from the alarming disequilibrium in the distribution of oceanographic capacity throughout the world.

According to UNCTAD and UNESCO figures, less than three percent of all funds committed to research and development in the world as a whole are actually spent in developing countries. Over ninety percent of all scientists and technologists live and work in developed countries. According to UNCTAD figures, the ratio of scientists and technologists per ten thousand inhabitants is ninety-five in the developing countries compared with 285.2 in the industrialized market economy countries and 308.2 in the Eastern European countries. Among the developing countries themselves, of course, there are significant differences. The situation in Asia, with 157.6 scientists per 10,000 inhabitants, is better than in Africa, with only 9.6 per ten thousand.

With regard to Research and Development and applied Research and Development, the situation is even worse. As summarized by Ivan Head in a recent article in *Foreign Affairs* (Winter 1989), the number of scientists, engineers and technicians engaged in Research and Development in the developing countries

is less than 1.5 per 10,000 inhabitants, compared with 16.6 in the market economies of the North.

Research and Development today is the prime engine of economic growth. According to some experts—e.g., M. Solow of the Massachusetts Institute of Technology—as much as eighty-five percent of U.S. economic growth per capita is attributable to increases in productivity due to technological innovation. Only about fifteen percent of the growth could be traced to the use of more inputs.

If we want to be serious about closing, or at least, narrowing, the gap between the industrialized and the developing countries, and building a new international order that is needed by the "North" as much as it is by the "South," it is here that we may find the real starting point. The advancement of the sciences and of applied Research and Development in cooperation with the developing countries, is fundamental for economic development, and the marine sciences may play a crucial role in this process.

If we want to be serious about banning the specter of nuclear holocaust and about enhancing military security together with economic and environmental security, it is here that we might make a break-through. Technologies developed through international cooperation can be used for peaceful purposes. In the absence of such cooperation, this development will remain restricted to a very few, very strong nations, which will pursue it through their military establishments for strategic purposes.

Thus a number of new factors converge to account for the increasing importance of oceanography all over the world: The growing contribution of the marine sector to the economy of coastal States (the growth rate of the marine sector is higher than the growth rate of global Gross National Product as a whole); the decisive role of new technologies in this process; and the direct linkage between these technologies and the marine sciences. Without the marine sciences, there can be no Research and Development; without Research and Development, no technology; without technological innovation, no development of the marine sector.

Oceanographic research, however, is extremely costly. Most developing countries cannot afford it on their own. Hence the need (and this is an additional, strong argument) for international cooperation, cost sharing, facility sharing, and generating economies of scale—both "South-South," on a regional basis, and "North-South," on a global basis. The 1982 United Nations Convention on the Law of the Sea encourages this kind of cooperation in many ways: through the establishment of regional centers of excellence; through cooperation with the competent international organizations, such as UNESCO and its Intergovernmental Oceanographic Commission, through bilateral assistance programs, training programs and personnel exchanges; through cooperation among oceanographers at the nongovernmental level—e.g., through the International Council of Scientific Unions; through international cooperative programs, many of which are described in this volume.

Apart from economic imperatives, however, it is the very nature of oceanography that necessitates international cooperation. For the nature of oceanography is determined by the nature of the medium in which it moves, the ocean, globe-spanning, where everything interacts with everything else. Fish do not recognize national boundaries, nor does pollution. The affliction of Arctic penguins by noxious chemicals produced thousands of miles away, noted in Chapter 3; the global climatic irregularities and disturbances generated by "El Niño," described in Chapter 10 and until recently deemed to be a local phenomenon, may illustrate the global scope of oceanography, in its investigations of this "turbulent fluid," in which "all space and time scales interact." The marine scientist is the prototype of a new interdisciplinary, international scholar, one who bridges the gap between theory and action. More often than not, the marine scientist is an explorer, a sailor, a diver, exposed to physical hardship and adventure, as well as a scholar, a theoretician, a philosopher of nature (see, for instance, Chapter 5 of this book).

Thus oceanography has become a household word in places where it was unknown in the past. A great many institutions in many countries have contributed, and are contributing, to our understanding of the ocean environment, its role in the global household of matter, of life, and of energy.

We have invited twelve of these institutions, on five continents, to present their "autobiographies," to describe their origins, their raison d'être, their structures and functions, their principal motivations and aspirations, their agendas for the future. Their work spans all the world's oceans, from the ice masses of the Arctic and Antarctic, which require their own technologies and their own type of scientist, to the fragile atolls and coral reefs of the tropical seas. They follow the great current systems linking all parts of the world ocean; they explore the mountain ranges and rift valleys running, through 40,000 miles, along the mid-ocean. They study the minerals and metals and their genesis in the submarine mountain mines. They study the ocean's fauna and flora and the origin of life. They measure the energy of tides and waves and define the ecology of coastal seas. They know that time and space dimensions differ near shore and in the open seas. They study the impact of human activities on the marine environment and the limits of the oceans' capability to absorb man-made waste. All life, and the air it breathes, originated in the oceans. Almost all waste, whether ship-borne or from land-based sources—human waste, agricultural waste, industrial waste—eventually ends up in the oceans, whether carried by rivers and run-offs or transported by the atmosphere and deposited through rain or snow. The furtive dumping of highly toxic or radioactive wastes in seas surrounded by poor developing countries has become a worldwide scandal. Recently a vessel, laden with 14,000 tons of incinerator ash, was trying to dump its cargo in the Indian Ocean. The vessel was refused entry into the waters of Sri Lanka, the Philippines, and Singapore. It later arrived in Malaysia without a cargo.

All the institutions portrayed in this volume contribute to the monitoring of the health of the oceans which is crucial for life on earth and the future of our planet. While there is a great deal they all have in common, each one of them also responds to the particular environment from which it emerged and within which it works: conditions determined by geography, by the kind of interests people have in the seas and oceans, their historic role as seafarers and explorers; the importance of the contribution of marine activities to the economic life of a nation, strategic interests, or national prestige. Each institution thus has a special contribution to make to the interdisciplinary, international effort of shaping the next phase of this incredibly exciting and fundamentally important science.

"To use the ocean wisely," E. Seibold wrote, "you must first understand it. How to maintain our renewable biological resources? How to conserve the genetic potential of the 180,000 marine species known up to now? How to conserve ecological integrity? What is the ultimate compatibility of the oceans with the many different kinds of pollutants transported from land by rivers and wind? All of these and many more problems need more and better research because at least since Bacon we have learned that 'Nature, to be commanded, must be obeyed'."

1.

The Scripps
Institution of
Oceanography,
California

Roger Revelle

Aerial view of the central part
of the Scripps Institution of
Oceanography, early 1980s.
Photo: © 1991 Wayne and
Karen Brown

When I first became a graduate student at the Scripps Institution of Oceanography in the fall of 1931, it was a rather puny place. Although it had been part of the University of California since 1912, its "campus" consisted of one laboratory building, a combined library and museum, a small aquarium building, several other little wooden buildings, and around twenty cottages for housing staff and summer visitors. There were also a thousand-foot pier and a sea wall. The whole enterprise was a field station of the University of California at Berkeley, one of whose regents had described it as consisting "mainly of a pier and a sea wall, both in need of repair."

The staff consisted of five faculty members including the Director, five graduate students, two spinster secretaries, a superintendent of buildings and grounds, an engineer, a gardener, a janitor, the curator of the small aquarium-museum, a couple of miscellaneous staff members, and a retired Canadian army captain who recorded tides and measured ocean temperatures at the end of the pier every day. One of the secretaries was also the librarian, the other doubled as the purchasing agent, storehouse keeper, accountant, personnel officer, and general business manager. We had one "ship," a retired purse-seine fishing vessel, named *Scripps*, sixty-four feet, ten inches long. (It had to be less than sixty-five feet long in order to qualify as a "small boat," which had minimum requirements for licensing.) Her single crew member was an ex-locomotive engineer who apparently believed that the best way to keep a boat in good shape was to cover it with grease like a steam engine. With this craft we were able to leave the port of San Diego for one- or two-week expeditions to the various islands and banks off the Southern California coast—San Clemente, Catalina, Santa Cruz, San Miguel, and San Nicholas among the islands, and Cortez and Tanner Banks—measuring ocean temperatures and other properties and collecting biological samples en route.

The Director was a nice old paleontologist whose sole sea-going scientific experience had been gained when he planted some young corals on wooden boards in the ocean off the Florida Keys to see how fast they would grow. The Institution was connected to the town of La Jolla by a three-mile-long two-lane road. The

The Scripps Institution of Oceanography in 1930: in the foreground, the library-museum; in the background, the George H. Scripps Laboratory; on the right, the wooden aquarium building. Photo: Passmore, San Diego, SIO Archives

Bridge and foredeck of the
Melville, night loading prior to
departure of expedition.
Photo: SIO Photo Lab

town's inhabitants usually referred to it as "The Biological" and sometimes as "the bug house." Most of the staff lived in wooden cottages on the institution grounds. The latter covered a total of 180, mostly unbuilt-on acres, on which there were supposed to be more than 150 different species of eucalyptus tree.

Today the Scripps Institution is the largest oceanographic institution in the world. It has 180 graduate students, 96 people with professorial titles (21 of them members of the National Academy of Sciences or the National Academy of Engineering, or both—there are also two Fellows of the Royal Society of London), about an equal number of research staff members, and a total of well over a thousand employees. It maintains four relatively large research ships and two sea-going research platforms, displacing all together about six thousand tons. At least two of these ships are always at sea somewhere in the world, or in an exotic port such as the Seychelles in the Indian Ocean or Tongatabu in the Pacific. In the last forty years vessels of the Scripps Institution have traveled over four million miles throughout all the oceans in a continuous series of voyages of exploration and discovery.

What happened?

Early Days

To answer this question we have to go back to 1903, when an adventurous University of California Professor of Zoology and an enthusiastic marine shell collector from San Diego collaborated to form the San Diego Marine Biological Association. The professor was William Emerson Ritter, head of the Zoology Department at Berkeley. For several years he had been carrying on a biological survey of the waters off the California coast. The enthusiastic collector was a practicing physician, Doctor Fred Baker. He was president of both the San Diego Board of Education and the County Medical Society, and he was a member of the San Diego Chamber of Commerce. In the spring of 1903 Dr. Baker wrote to Professor Ritter to ask him whether he would be willing to bring his biological investigation to San Diego for the summer, if interested local people could raise the necessary funds. Ritter replied: If Baker and his friends would locate and equip a suitable laboratory, and in addition provide five hundred dollars for operating expenses, the University of California's Department of Zoology would carry on its activities in the summer of 1903 at San Diego. With the help of a special marine laboratory committee set up by the San Diego Chamber of Commerce, Baker raised

$1,300. He arranged with the manager of the Hotel Del Coronado in San Diego Bay to use their boathouse as a summer laboratory. Also, he recruited one of his neighbors, a Portuguese fisherman named Manuel Cabral, to make biological collections for the scientists, using a small chartered schooner. But the most important thing he did was to introduce Ritter to E. W. Scripps, the great newspaper man, and his sister, Ellen Browning Scripps.

Ritter and E.W. took to each other immediately. They were both idea men who liked nothing better than to talk and plan for the future. For most people, E.W. was a nonstop loud talker with whom it was hard to get a word in edgewise. But he and Ritter actually were able to carry on a two-way conversation, which they did at very frequent intervals for the next twenty years.

The most important task that E.W. accomplished for the fledgling institution was to talk the City of San Diego into selling one hundred and eighty acres of land to the Marine Biological Association, for a thousand dollars. This has now become the southernmost part of the campus of the University of California at San Diego. It contains the ten relatively large buildings, covering several hundred thousand square feet, and many of the numerous small structures that now house most of the Scripps Institution of Oceanography. Other laboratories and shop buildings and the land-based part of the Scripps ship facilities are located in Point Loma on the shores of San Diego Bay. Part of the Institution is under water as a national marine biological reserve.

E.W.'s sister, Ellen Browning Scripps, became the principal benefactress of the Marine Biological Association. Year after year, until the early 1920s, she provided several tens of thousands of dollars annually for operating expenses, and in the annus mirabilis of 1915 she gave a hundred thousand dollars to build a pier, a sea wall, and several buildings, one of which is still standing as a "historic monument" of a laboratory.

The Marine Biological Association of San Diego was incorporated in 1904, with regular by-laws, an established board of directors, and provisions for financial accountability. E.W. Scripps had insisted on this. He said that before it was incorporated the association "conducted its affairs by the parliamentary rules of a tea-party." It became part of the University of California in 1912 as the Scripps Institution for Biological Research. Note that the word "Marine" was dropped. One reason for eliminating it was E.W.'s and Ritter's growing interest in doing research on what they called "philosophical biology." Another reason was the presence of tens of thousands of mice in the

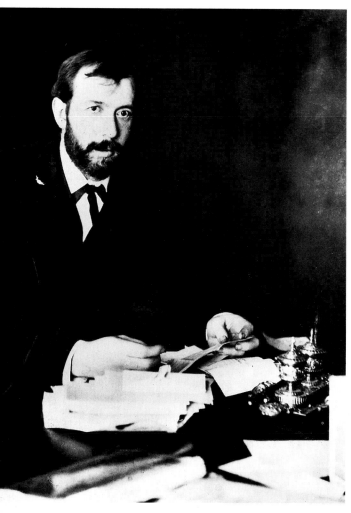

Edward Willis Scripps, 1903,
about the time he and his
sister, Ellen Browning Scripps,
first became interested in
Professor Ritter's project.
Photo: © 1903 by Cecil's
House of Photography, SIO
Archives

Institution. They were being used by Professor F.B. Sumner to test Lamarck's hypothesis of the inheritance of acquired characteristics. Professor Sumner watched his mice breed for many generations, but he never found an acquired characteristic that was inherited. For this and his (earlier and later) work on the way in which halibut and other flat fish adjust their skin color patterns to match their environment, he was elected to the National Academy of Sciences in the late 1930s.

After the Scripps Institution became part of the University of California, Dr. Ritter and his colleagues had to go directly to the legislature for funds. This would undoubtedly shock the present University administration, which insists that only the office of the president of the entire University of California should approach the legislature. Judging by the results, the president and his staff have the better of the argument. Dr. Ritter was able to obtain appropriations of $12,900 in 1915 and 1916, and $17,500 in 1919 and 1920. In 1988 the legislature appropriated $15,630,000 to the Scripps Institution, a thousand times the appropriations Ritter was able to squeeze out of them.

In 1925 the Institution experienced its final change in name. It became the Scripps Institution of Oceanography. The reason was that Ritter had become very dissatisfied with the reductionist trend in biology that had begun a few years before under the leadership of his colleague at Berkeley, the great Jacques Loeb. Biological research was moving from the study of whole animals and plants to the study of organs, and then to cells, and finally to the molecules that make up the cells. The giant strides biology has made during this century have depended partly on this trend in research toward ever lower levels of biological organization. But Ritter thought it was wrong. He wrote a famous book, entitled *The California Woodpecker and I*, in which he forcefully argued his view that the proper objects of biological research were whole organisms and their relationships with their environment. He envisioned that with its new name and aim the Scripps Institution would concentrate on these relationships in the ocean realm.

It was an oceanographic institution without a seagoing ship, though it did have a pier—still a small institution, very much underdeveloped, even for carrying out Dr. Ritter's plan of a biological survey of the waters off Southern California. But he had great hopes and grand plans for it. He said, "One cannot adopt a baby elephant for a pet without sooner or later having a big elephant on his hands if he treats the creature humanely." He was not interested in size. "If the station does the things I would have it do it will unfortunately

Scripps' Nimitz Marine Facility
on Point Loma in San Diego
Bay. Photo: SIO Photo Lab

have to be rather big. But my ambition for it—and in this I am sure I speak for its patrons and official friends as well as for myself—is that it should be great rather than big." The patrons he referred to were, of course, E.W. Scripps and Ellen Browning Scripps.

It seems to be a requirement that to be Director of the Scripps Institution one has to be a visionary with grandiose ideas and implausible plans. When I was Director, for example, I told the University's regents that the biggest county in California was Pacific Ocean County—the entire Pacific Ocean, which I urged them to incorporate as a major resource of the state of California.

The Great Transformation

The transformation of the Scripps Institution into a modern, world-ranging, broadly-based center of research and teaching in the earth and ecological sciences began in 1936 when Harald Ulrik Sverdrup became Director, and our little boat, *Scripps*, blew up in San Diego Bay.

Sverdrup was a world-class geophysicist, a leading member of the school of ocean and atmospheric scientists that grew up in Norway, mostly in the small city of Bergen, before World War II. He had spent seven years, in two successive voyages, frozen into the Arctic ice with six fellow Scandinavians on Roald Amundsen's ship, *Maud*. He was a small, dark man, not at all your classical large, blond Scandinavian, but nevertheless typical of one kind of Norwegian one finds in the Western fjords of Norway. His experience in the ice had both toughened and gentled him, and given him a rare level of insight into the strengths and weaknesses of men.

After Harald Sverdrup had been in La Jolla a few months the little old ex-purse-seiner, *Scripps*, destroyed herself in an explosion, which also killed the cook (by this time the boat had acquired a crew of two) and seriously injured the engineer. Despite this tragedy, the accident was ultimately a blessing for the Scripps Institution. It enabled Sverdrup to approach E.W. Scripps' son, Robert Scripps, with a request that he provide the Institution with a real seagoing vessel. This was done, and the Institution came into possession of a beautiful topsail schooner, which was renamed E.W. *Scripps*.

In his seven years in the ice Sverdrup lost most romantic notions about sailing ships, and he promptly had E.W. *Scripps'* topmasts removed and a much larger engine installed. After this modification we were able

to carry out two long expeditions to the Gulf of California and to conduct several systematic surveys of the currents and water masses off the coasts of California, Oregon, and Baja California.

World War II came in the midst of these activities. Most of the Scripps staff became involved in the war effort, either in a uniform, or at the temporary University of California Division of War Research (UCDWR) on Point Loma in San Diego. Its research and development was concentrated very largely on the technical problems of submarine and antisubmarine warfare. These turned out to have a strong oceanographic component—a fact that was discovered more or less simultaneously by the Woods Hole Oceanographic Institution in the Western Atlantic, and by the British Admiralty in the East. Sverdrup and his student, Walter Munk, concentrated on developing a forecasting system for ocean waves and swell and surf in shallow water. They taught this system to several classes of young Army and Navy officer meteorologists; it was used to forecast breakers and surf on the landing beaches of Sicily and Normandy in the Atlantic and on many island beaches in the Pacific. One major result of these exercises was that the American and British Navies discovered the multiple values of oceanography for naval and amphibious operations.

After the war's end this discovery plus the establishment of the Office of Naval Research, which took a liberal, farsighted view of the ways in which the United States Government could and should support basic research, led to a virtual revolution in the magnitude and quality of the funding available for Woods Hole and Scripps. And in a few years it led to the establishment or transformation of a series of oceanographic stations on all the coasts of the United States.

Using two seagoing tugs and a wooden minesweeper provided by the Navy, the Scripps Institution expanded in two directions.

In his last two years as Director of Scripps, from 1946 to 1948, Dr. Sverdrup planned and organized a large-scale oceanographic investigation. In cooperation with the United States Bureau of Commercial Fisheries, the California Department of Fish and Game, and the Hopkins Marine Station of Stanford University, Scripps began a long-term continuing study of the oceanographic and biological conditions off the California and Baja California coasts that were believed to affect the size, location, and growth of commercially important pelagic fish populations. These "CALCOFI" studies (California Cooperative Oceanic Fishery Investigations) are still continuing. They constitute an unequaled record of environmental stability and change

Harald Ulrik Sverdrup, Scripps
Director from 1936 to 1948,
adjusts a current meter on
board E.W. *Scripps*. Photo:
Eugene LaFond, SIO Archives

Schooner E.W. *Scripps*, about
1939, the Scripps Institution
first deep sea research
vessel—a gift from Robert P.
Scripps, youngest son of E.W.
Scripps. Photo: SIO Archives

and their biological consequences in a large oceanic area. During the forty-year period the population of California sardines (*Sardinops caerulea*) virtually disappeared and their place was taken by the Northern anchovy (*Engraulis mordax*). Now the sardines seem to be returning and replacing the anchovies. This ebb and flow between the two species has apparently occurred several times in the past, judging by fossil fish remains in varved sediments of the Santa Cruz Basin south of Santa Barbara. These sediments were deposited before the incursion of human fishermen into California waters.

The New Age of Exploration

The other change in the direction of Scripps activities began in 1950 with the Mid-Pacific Expedition to equatorial waters and to the Marshall Islands in the Central Pacific. This was a two-ship expedition, conducted jointly with the United States Navy Electronics Laboratory in San Diego. It resulted in a set of remarkable discoveries about the ocean floor and what lies beneath it, and was the first of a long series of expeditions, extending further and further from San Diego until the Scripps ships literally operated in all oceans throughout the world.

The most important discoveries of the Mid-Pacific Expedition were the demonstration by Russell Raitt and his associates that only a thin layer of deep-sea sediments one hundred or two hundred meters thick overlies solid rock, presumably volcanic lava in the North Pacific, and the discovery by Edwin Hamilton of fossil shallow-water corals about sixty-five million years old on top of a flat-topped sea mount, or guyot, at a depth of two thousand meters. This showed that the guyot had been a volcanic island which was planed off just below sea level by wave action during the latter stages of the Age of Dinosaurs. The guyot was a peak in a submerged mountain range, extending from Necker Island in the Hawaiian Chain to Wake Island in the Northwestern Pacific. We called this newly discovered mountain chain the Mid-Pacific Mountains. Previously, the guyots had been thought to be many hundreds of millions of years old, and their depths of several thousand meters or more had been explained by the filling up of the ocean basins with several thousands of meters of marine sediments.

On the Mid-Pacific Expedition Arthur Maxwell, a Scripps graduate student (who later became Director of the Geophysical Institute of the University of Texas) measured for the first time the heat flowing from the interior of the Earth through the seafloor. Much to the surprise of all of us, the heat flow was about equal to that coming from deep mines on the continents. Later, after Art had made more measurements across the Mid-Ocean Rise in the South Pacific to the South American Trench, he and I and Sir Edward Bullard showed that these heat flow measurements could be accounted for only if slow convective movements were occurring in the mantle of the Earth. Unfortunately, we did not suggest that this convection was dragging the seafloor with it from the Mid-Ocean Ridge to the Trench, where it sank deep into the interior, thereby causing a constant replacement of old seafloor with new material. This phenomenon of "seafloor spreading," first suggested by Harry Hess of Princeton University and Robert Dietz, a former Scripps graduate student, of course accounts for the thin layer of sediments discovered by Russell Raitt.

The thinness of the deep-sea sediments was confirmed in an extended voyage to the South Pacific in 1952–53, which we called the Capricorn Expedition. One of its objectives was a study of the deep-sea trench east of the Tonga Islands which extends from a region south of Samoa for 1,500 miles almost to New Zealand. It turned out to be over 35,000 feet deep, the deepest spot in the Southern Hemisphere, only a few hundred feet shallower than the great deep east of Guam, that is farther below sea level than any place in the world. When we tried to core or dredge in the trench the instruments came up battered and bent, and empty. If there were any sediments in the bottom of the trench, they were sparse and thin. On its eastern flank we found a guyot, a flat-topped sea mount like those in the Mid-Pacific Mountains. But its summit was not level—it was dipping toward the center of the trench. These observations could best be explained if the rocky seafloor was disappearing into the earth along the axis of the trench. This process, which is now called subduction, also accounts for the deep epicenters of earthquakes along a steeply dipping plane beneath the landward side of the Tonga Trench and the other trenches that ring the Pacific.

One of the instruments used on the Capricorn Expedition was a device for measuring the earth's magnetic field from a moving airplane or ship. It had been invented by Victor Vacquier, who later became a Scripps staff member. On Capricorn it was the special charge of a young English geophysicist, Ronald Mason. As the ship moved along, the magnetometer was towed about a hundred feet astern. Ronald was able to record a complicated set of wiggles which none of us understood very well. After the return of the expedi-

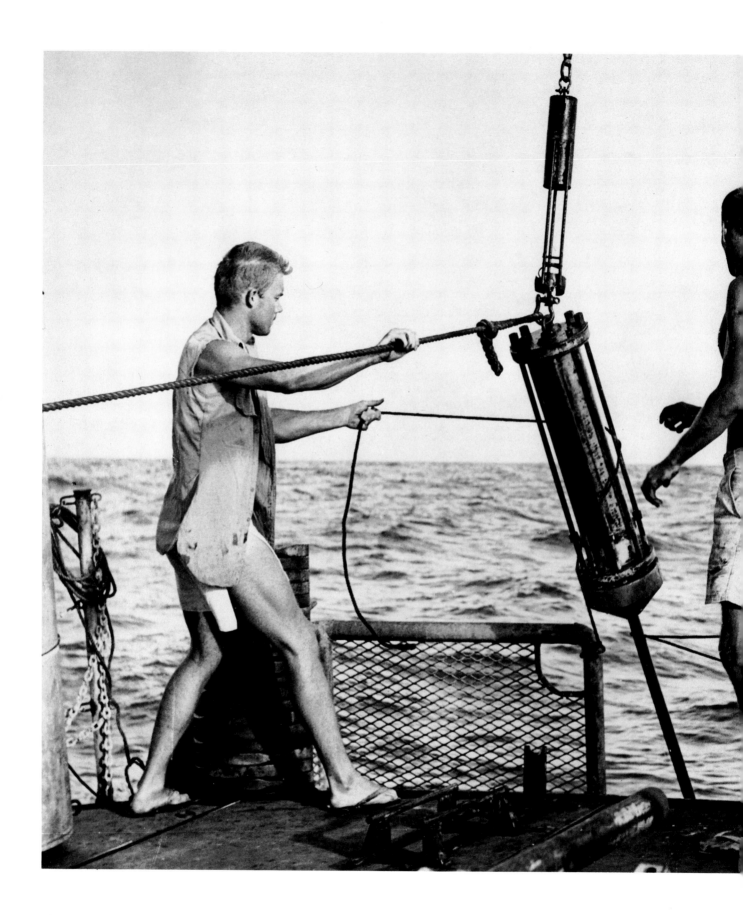

Art Maxwell, Scripps graduate student, and assistants putting the temperature probe overside on the Scripps Capricorn Expedition of 1952–53 to the South and Central Pacific. Photo: SIO Archives

tion, he managed to persuade the captain of the United States Coast and Geodetic Survey ship, *Pioneer*, to let him tow his magnetometer during a detailed survey of the bottom topography off the California coast. The results were spectacular. He produced a map of the magnetic field under the seafloor that showed long lines of alternating magnetic ridges and troughs, completely unrelated to the sea bottom topography, except in the neighborhood of a sea mount. It was later shown that the magnetic ridges and troughs are symmetrically arranged on the two sides of a mid-ocean ridge. Each ridge and trough corresponds to a reversal of the earth's magnetic field, which takes place at intervals of somewhat less than a million years, and hence it is possible to date the times when volcanic lavas upwelling along the mid-ocean ridges formed different portions of the ocean crust. It turns out that the mean age of the ocean floor is only about a hundred million years.

The final confirmation of the hypothesis of seafloor spreading came two decades later, when Arthur Maxwell and his associates were able to drill along a section from the center of the mid-Atlantic ridge to the continental shelf off Brazil. They found only relatively young sediments near the axis of the ridge, and progressively thicker sediments of increasing age-span, up to fifty million years, as they approached the Brazilian coast. Sediments or bottom rocks older than about a hundred and fifty million years have not been found anywhere in the ocean. In contrast to the continents, which in places contain rocks one to three billion years old, the seafloor is young and is continually renewed.

These discoveries about the seafloor, together with paleomagnetic measurements and paleoclimatic evidence first obtained by British scientists that demonstrate that the continents have moved for great distances over the earth's surface, are the scientific underpinning for the theory of plate tectonics. This theory has literally revolutionized our understanding of the history of the earth, including the ways in which plants and animals have evolved and been distributed over the oceans and continents.

In those heady days of the 1950s one could hardly go to sea without making an important, unanticipated discovery. Our small ships didn't cost very much to operate and many SIO expeditions were led by graduate students. John Knauss, former Dean of the School of Oceanography at the University of Rhode Island, and now administrator of NOAA, obtained his Scripps Ph.D. on the basis of expeditions to study the equatorial undercurrent in the Pacific. This powerful, east-

ward flowing current had only recently been discovered by another graduate student, the late Townsend Cromwell. Warren Wooster, past president of the International Council for the Exploration of the Sea, led several expeditions when he was a graduate student at Scripps to study the currents and water masses of the Central North Pacific. Robert L. Fisher, still on the Scripps research staff, surveyed the deep trench off Central America for his thesis and showed that it was probably moribund, no longer being subducted. We have already discussed Arthur Maxwell's heat flow measurements which he made and used for a Ph.D. thesis after leading an expedition in the Southeastern Pacific.

Scripps Gives Birth to a University and Becomes Organized

Another great transition in the life of the Scripps Institution began in 1960, when it gave birth to a major undergraduate and graduate campus of the University of California. This academic infant has now grown to giant size compared to its parent. The original 180 acres of the Scripps campus has been consolidated in an area of 1,200 acres which is gradually filling up with students and faculty and all the manifold activities of a modern "research university." The Director of the Institution has now gained two additional titles. He is the Dean of Marine Sciences, because Scripps is a graduate school, and he is Vice-Chancellor of Marine Sciences, to indicate the still important role that the Scripps Institution plays in the University.

The Scripps Institution has become both broader in its scientific concerns and more complicated in its organizational structure. In its teaching program it is consolidated into a single university department, but with seven divisions: Applied Ocean Sciences, Biological Oceanography, Geochemistry and Marine Chemistry, Geological Sciences, Geophysics, Marine Biology, and Physical Oceanography. A student may pursue a five- or six-year Ph.D. degree in any one of these divisions.

Over the three-year period 1986–1988, seventy-seven Ph.D.s were awarded in Applied Ocean Sciences—the curriculum combines Scripps resources with those of two engineering departments in the University. In Biological Oceanography the students study the interactions of marine organisms with the oceanic environment and with each other. The Geochemistry and Marine Chemistry curriculum emphasizes chemical and geochemical processes in the ocean, the atmo-

Walter Munk, Scripps graduate
student, looking at a giant
coral at a depth of thirty
meters on Alexa Bank in the
South Pacific, during the
Scripps Capricorn Expedition
of 1952–53. Photo: SIO
Archives

Scripps' Mark Zumberge and
colleagues towing a sled-
mounted Global Positioning
System instrument on the
Greenland ice cap to locate
positions for an areal gravity
survey. Photo: Glenn
Sasegawa

sphere, lakes, rivers and glaciers, meteorites and the solid earth. The Geological Sciences curriculum applies observational, experimental, and theoretical methods to the understanding of the solid earth and solar system and how they relate to the ocean and atmosphere. The Geophysics curriculum is designed to educate physicists about the sea, the solid earth over which the waters move, and the atmosphere with which the sea interacts. The Marine Biology curriculum encompasses a range of biological disciplines, including behavior, neurobiology, developmental biology, and comparative physiology/biochemistry. Studies in Physical Oceanography include observations, analysis, and theoretical interpretation of the general circulation of ocean currents, transport of dissolved and suspended substances and heat, the propagation of sound and electromagnetic energy in the ocean, and the properties and propagation of ocean waves.

In its research activities the Scripps Institution operates through three divisions—Geological Research, Marine Biology, and Ocean Research; two special research units—the Center for Coastal Studies and the Marine Life Research Group; two laboratories: the Marine Physical Laboratory and the Physiological Research Laboratory; a neurobiology unit; and three University of California-wide institutes: the California Space Institute, the Institute of Geophysics and Planetary Physics, and the Institute of Marine Resources. The latter contains the Sea Grant College Program, which is analogous in many respects to the University of California's activities as a Federal land grant college.

Technical and administrative support for these research and teaching programs is provided by the ship operations and marine technical support staff; by the existence of four scientific collections, each with its own curators (benthic invertebrates, geological samples, marine vertebrates, and planktonic invertebrates); and also by the Scripps Library; the Satellite Oceanography Facility; a computing facility which, among other services, provides access to the San Diego Supercomputer Center; and the Thomas Wayland Vaughan Aquarium-Museum, named after the second Director of the Institution.

The Scripps Library currently receives more than 3,800 serial titles, and contains over 200,000 volumes, including an extensive collection of technical reports and translations and a rare books collection featuring accounts and journals of famous voyages of scientific discovery. A large map collection contains bathymetric, geologic, and topographic maps and charts of world areas and oceans. The library also houses the archives of the Scripps Institution of Oceanography, which include official Scripps records, personal papers, photographs, and other material documenting the history of oceanography and of Scripps.

The Scripps Satellite Oceanography Facility enables oceanographers to receive and process satellite imagery, including data transmitted in real time by the NOAA Polar Orbiting Satellites.

Among other physical facilities are two research aquariums, a holding pool for large marine mammals and fish, a ring pool of ten meters radius and a seawater system which pumps seawater from the seaward end of the pier, providing up to 6,500 liters per minute to the public aquarium and the research laboratories, as well as to the adjacent Federal National Marine Fisheries Laboratory. The Institution operates radio station WWD which provides worldwide communications services to Scripps and other University and governmental scientific ships. It also conducts a scientific diver training and certification program. After a hundred hours training in the use of self-contained underwater breathing equipment (SCUBA) an average of 130 students a year receive a University of California Research Diving Certificate.

The Scripps Institution operates two remote measuring facilities. The Cecil and Ida Green Piñon Flat Observatory contains a group of highly precise geodetic instruments to measure earth tides and other motions in the earth's crust. The San Vicente Lake Calibration Facility is used for testing and calibrating acoustic transducers designed for oceanographic research. It is located in water forty meters deep and offers an unobstructed range of 1,372 meters.

What can be said about the research products of the faculty and staff and their supporting facilities? In quantitative terms we can count the average 300 scientific papers and reports by some 500 authors published annually. Supposing half the costs of operation of the Institution can be charged against its teaching activities and half against research, each publication costs a little over one hundred thousand dollars.

But a more important measure is the advances in understanding of the earth and its creatures that have resulted from Scripps Institution research. We have already told about some of these in our story of the early expeditions. We can go further by discussing the work in different fields—biology, geochemistry, physical oceanography, geology, and geophysics.

Biological Research at Scripps

The contributions of SIO biologists to man's understanding of marine life span the entire spectrum of the

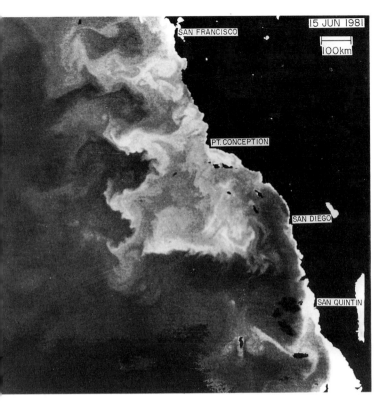

Satellite image of chlorophyll in the California current in June 1981. Light gray shades correspond to relatively high chlorophyll content (high levels of phytoplankton photosynthesis). Dark gray corresponds to low chlorophyll. Note sharp boundary southwest of San Diego and eddies on the western side of the current. Image processed by J. Pelaez of the Scripps Satellite Facility

biological sciences, from broad, field ecological studies to biochemical and molecular analyses. During the Institution's early years, its biologists focused on the description of the types of organisms found in the intertidal, near-shore and open ocean environments. This tradition, the taxonomic identification of new species and the recording of their numbers and distributions, continues today. SIO maintains an active curatorial program in fishes and marine invertebrates that provides a key basis for the experimental biology at La Jolla and, indeed, at marine institutions worldwide. The knowledge available from the curatorial staff and its collections provides a data base of "what's out there," allowing field scientists to conduct meaningful studies of natural variations in populations.

The discovery by Martin Johnson and others during World War II of the true nature of the *deep scattering layer* certainly merits a place near the top of a list of important research accomplishments in marine biology. The demonstration that a large and complex assemblage of marine animals moves vertically through the water column on a diurnal basis has led to numerous important lines of marine biological research, e.g., on the composition of this group of species and on the energetic implications of vertical migration in the sea.

The work of Claude ZoBell on deep-sea bacteria, work which has been continued and significantly extended by Aristides Yayanos, demonstrated that bacteria from the deep sea are true *barophiles*—species which not only tolerate elevated hydrostatic pressures but which, in fact, can survive only under high pressure. ZoBell and Yayanos thus proved that there is a unique bacterial population in the deep sea. Yayanos has shown that the pressure tolerances and pressure requirements of marine bacteria increase in a fairly uniform manner with increasing depth in the marine water column. Yayanos also showed that very deep-living marine animals are also barophilic, requiring elevated pressures for their survival. Molecular level studies by George Somero's group showed that adaptation to the deep sea entails the modification of numerous protein systems, including metabolic enzymes and the proteins that drive muscle contraction. Even pressures corresponding to depths of 500 to 1,000 meters favor the evolution of new protein variants with pressure-adapted characteristics.

Similar adaptations of marine organisms to different temperatures were studied by one of Scripps' great biologists, the late Carl Hubbs. He demonstrated, by his studies off Baja California and Central America that discontinuities in species distribution are often associated with discontinuities in ocean temperatures. These observations can be at least partially explained by Somero's finding that temperature differences of the order of five degrees Celsius favor evolutionary selection of different protein forms. His studies provide a link between ecological and molecular properties of marine organisms, helping to explain why species thrive in—and usually only in—the temperature and pressure regimes in which they have evolved.

Working at more ordinary pressures and temperatures, Farooq Azam and his co-workers have made major contributions to the understanding of the role of bacteria in marine food webs. They have shown that bacteria may be of primary importance in the cycling of organic matter in certain marine habitats. Marine bacteria have the abilities to locate food particles, to produce the enzymes needed to hydrolyze these particles, and to transport into their cells the solubilized nutrients released by this hydrolysis.

Osmund Holm-Hansen has spearheaded Scripps biological and biochemical research in the Arctic and Antarctic Oceans, using modern molecular-biochemical techniques to study particularly phytoplankton adaptation to light and the quantum efficiency of photosynthesis. The large biomasses in these regions make them increasingly important as potential food resources.

Perhaps the most widely known biological work done by SIO scientists in the deep sea is the study of hydrothermal vent communities. After these communities were discovered at the Galapagos Spreading Center in 1977, SIO biologists made many of the key contributions to our understanding of the types of organisms found at these sites, and of how they succeed in an environment which, on several grounds, seems very "hostile." The discovery that the large vent tube worm (Riftia pachyptila) houses within its body a large population of sulfide oxidizing bacteria which can use the energy released in the oxidation of sulfides to drive carbon fixation via the Calvin-Benson cycle (the same pathway used by green plants in photosynthesis) was made by Dr. Horst Felbeck of Scripps. Subsequent work on these animal-sulfur bacteria symbioses by Dr. Mark Powell, George Somero, and Russell Vetter of Scripps showed that both the animal and the bacterial components of the symbiosis can exploit the energy of the sulfide molecule. This was the first demonstration that animals can use reduced inorganic compounds as foodstuffs.

Work with marine mammals has been an important focus of scientists in Scripps' Physiological Research Laboratory. Dr. Per Scholander, who established this Laboratory, discovered the diving response of seals and some other marine mammals. When placed under water, the animal changes its heart rate and circulatory patterns to conserve oxygen. Later work by Dr. Gerald Kooyman showed that Antarctic Weddell seals typically dive for periods of less than one-half hour, a period during which their metabolism can remain aerobic. Only under extreme cases, i.e., during very long dives, does the animal need to rely on anaerobic metabolism to drive its swimming muscles. The situation is quite different with elephant seals, which typically dive to more than fifteen hundred meters and remain at depth for extended periods. Kooyman's field studies of marine mammals are classic investigations which have combined sophisticated in situ measurements with precise laboratory analysis of blood and tissue biochemistry.

California Sea Grant

The National Sea Grant College Program is similar in intent and function to the 125-year-old College Land Grant Program, which provides Federal support to universities in every state for cooperative projects in agricultural research and extension. The headquarters of the California Sea Grant College Program are in the Scripps Institution of Oceanography, but all campuses of the University of California participate in the program together with ten campuses of the State University system, several community colleges, and a group of private institutions. Space does not permit a description of the many activities of this program—developing mechanisms for coastal zone management, manipulating coastal wetlands, studying physical processes in the coastal zone, ocean engineering, and ocean technology. We shall concentrate on some aspects of the work in fisheries biology and aquaculture and in marine pharmacology.

For many years salmon hatcheries have been operated on both sides of the Atlantic and Pacific Oceans in efforts to increase the commercial yields of ocean-run mature salmon. A complex physiological process called "smoltification" occurs when the young fish metamorphose into seawater-adapted "smolts" before beginning their journey to the sea. Sea Grant researchers found that the young fish exhibit a surge of thyroid activity at the time of metamorphosis, and they were able to show that this thyroxin surge coincides with the new moon following the spring equinox. Fish released from hatcheries at this time show up two

Giant tube worms (up to 2.6
meters long) swaying in
waters near a hydrothermal
vent in the Galapagos
Spreading Center. Photo: Dr.
James S. Childress, University
of California, Santa Barbara,
© Scripps Institution of
Oceanography—USCD

Deep-sea hydrothermal-vent
fish, Bythites, n. sp., captured
by Woods Hole Oceanographic
Institution research
submersible, Alvin, in June
1988, during a joint Scripps-
Woods Hole expedition to the
Galapagos Islands Spreading
Center.

Roger Revelle, Scripps
Director from 1951 to 1964,
visiting Francis Haxo's
laboratory of marine plant
physiology, about 1955. Photo:
© 1958, SIO Archives

or more years later in far greater numbers in the fishery, clearly implying greater survival. The ability to predict migratory readiness by the lunar calendar has minimized the need for blood sampling and complicated technology.

In the late 1880s the United States was the world's second largest producer of caviar. But American sturgeon were soon destroyed by overfishing and destruction of habitat. Today serious efforts are being made to domesticate sturgeon. This work began in 1977 with California Sea Grant support in cooperation with Federal and State agencies, and today there are twelve sturgeon hatcheries and ten growers. They routinely spawn male sturgeon, but they must still capture mature females which, in nature, take fifteen to twenty years to mature. Bringing cultured females to maturity is now a top priority. In the meantime, California aquaculturists are marketing fingerling sturgeon to other growers in the United States and Europe and to the aquarium trade, and the fish are being planted in lakes for recreational fishing. Sturgeon farming could become a major aquaculture success story in future years.

Among marine invertebrates, the sea urchin fishery has rapidly become California's largest and most valuable shellfish fishery. Landings have tripled since 1985. Sea urchin roe are a great delicacy in Japan; exports from California to Japan help a little in balancing our trade deficit. In cooperation with scientists of the National Marine Fishery Service, Paul Dayton and Mia Tegner of Scripps and Joseph Connell of the University of California at Santa Barbara have studied the sea urchin's role in near shore ecology, especially in the kelp beds off the Southern California coast. Their work has been of critical importance in managing the urchin fishery.

Another development in marine aquaculture has concerned attempts to culture abalones, a much prized catch in California, which have suffered sharp declines in population and commercial landings since the early 1960s. A first problem faced by Sea Grant researchers was that of inducing spawning. It was found that spawning could be controlled by adding small quantities of hydrogen peroxide to the seawater. The second problem was to induce the free swimming abalone larvae to undergo metamorphosis and to settle on a rock surface. Daniel E. Morse and his associates in the Sea Grant Program at the University of California, at Santa Barbara, found that gamma aminobutyric acid could be used to induce larval metamorphosis and settlement, and this substance is being utilized on a routine basis by one commercial abalone company in

California and by others in New Zealand. Unfortunately, attempts to plant hatchery-reared juveniles in the open ocean have largely failed, because the hatchery-reared animals tend to settle on exposed sides of bottom rocks rather than underneath the rocks where they would be hidden. Much greater success has been attained in culturing the Pacific oyster, *Crassostrea gigas*, with the help of California Sea Grant researchers.

Marine plants and animals are a genetically unique resource with vast potential for providing pharmaceuticals, agrichemicals, and industrial chemicals of novel types. Several campuses of the University of California have cooperated in California Sea Grant research on these substances. Under the leadership of William Fenical and John Faulkner of Scripps, Phillip Crews of Santa Cruz, and Robert Jacobs of Santa Barbara, working with Isao Kubo and Neylan Vedros in Berkeley and James Sims in Riverside, a marine pharmacology program has been established. By 1985 over 600 compounds from various marine plants and animals had been isolated and purified. At the beginning of the program in the early 1970s, it was shown that marine plants contain organohalogen compounds that are very similar in structure to such halogenated pesticides as DDT and possess equally powerful insecticidal properties. A few years later, red seaweeds from California coastal waters were found to contain compounds with powerful antiviral properties, capable of controlling herpes simplex, for example. In the early 1980s an unprecedented new class of antiinflammatory agents was discovered based upon the structure of manoalide, a compound isolated from a tropical sponge. This compound acts by a mechanism that is unique among all anti-inflammatory agents. More recently Crews and his colleagues have collaborated with the Syntex Corporation in developing two new classes of antifungal and antihelminthic drugs. Another unique anti-inflammatory and analgesic agent has been isolated from a Caribbean soft coral.

Geochemistry at Scripps

Perhaps the most widely known chemical studies at the Scripps Institution are Charles D. Keeling's measurements of the secular increase of carbon dioxide in the atmosphere, beginning in 1958, and continuing up to the present in cooperation with the National Ocean and Atmospheric Administration. During this period the carbon dioxide content of the air, measured near the top of Mauna Loa, the great volcanic peak on the Island of Hawaii, has increased from 315 ppm by vol-

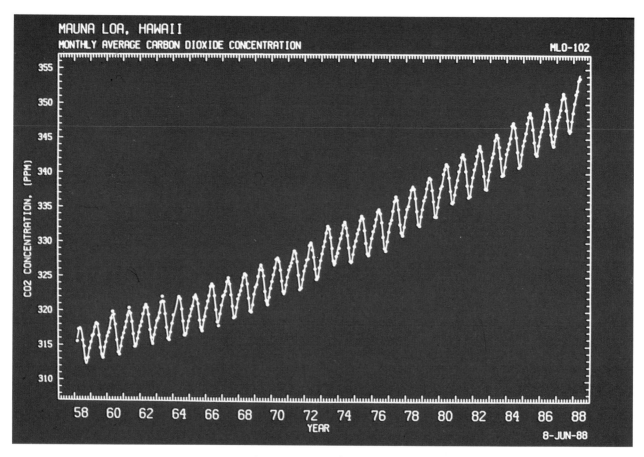

Continuous record of
atmospheric carbon dioxide at
the top of Mauna Loa on the
Island of Hawaii, measured by
Scripps' Charles David
Keeling, 1958–88.

ume to over 355 ppm, averaging somewhat more than 1 ppm/yr, or nearly thirteen percent in thirty-two years. The rate of increase is itself increasing and is now nearly one percent a year. A similar increase is found at the other observing station established by Dr. Keeling, at the South Pole.

Before Keeling began his measurements many people had suspected that atmospheric carbon dioxide was increasing—indeed Svante Arrhenius in 1898 wrote about mankind "evaporating our coal beds." But no one knew the rate of increase. All previous measurements had been taken near cities or forested regions where the atmospheric CO_2 content is likely to vary from season to season by more than twenty-five percent. It was generally thought, however, that the increase in the atmosphere must be very slow, because most of the CO_2 being released by fossil fuel combustion was probably being absorbed in the ocean. The oceans contain about sixty times as much CO_2 as the atmosphere, and the general belief was that CO_2 from fossil fuels would be partitioned between the sea and the air in the same ratio. Computations by Hans Suess and me indicated, however, that about half the fossil fuel carbon dioxide would remain in the air and less than half would be absorbed by the ocean. (We thought that the remaining CO_2 would be taken up by forests and other land biota.) After a few years, Keeling's measurements, compared with worldwide data on oil, natural gas, and coal consumption, showed that our predictions were correct, and a world "cottage industry" of atmospheric CO_2 measurements has developed on islands and coastal stations from the far Arctic to Antarctic in all three oceans.

One of the interesting aspects of these continuous CO_2 measurements is the annual oscillation between summer and winter. (The oscillation is highest in the far north and very small in the oceanic regions of the southern hemisphere.) In summer, plant photosynthesis exceeds biological respiration from plants, animals and bacteria, and CO_2 is taken out of the air. In winter, just the reverse occurs, and atmospheric CO_2 rises. The amplitude of this seasonal oscillation is increasing, probably indicating that forests and other biota are growing because of the fertilizing effect of increased atmospheric carbon dioxide.

Since Keeling began his measurements it has been realized that other gases besides carbon dioxide are "greenhouse gases," i.e., they act to absorb and back-radiate infrared radiation and thereby to heat the lower air. These other greenhouse gases include methane (natural gas), nitrous oxide, tropospheric ozone, and the "freons" or chlorofluorocarbons. They are all present in much smaller concentrations than CO_2, but molecule for molecule they have a much larger heating effect. Hence methane, for example, though it is present to the extent of only a few parts per million in the atmosphere is expected to account for perhaps twenty-five percent of the total greenhouse heating effect after a few decades.

An understanding of the sources of methane is therefore essential in efforts to mitigate the greenhouse effect. From analyses of cores of glacial ice, Harmon Craig and his colleagues found that atmospheric methane began to increase in Shakespeare's time, and by the beginning of the 1960s its concentration had doubled. It is now increasing at more than one percent a year; measurements of carbon isotopes in methane show that about twenty percent of this increase originates in leakage from oil and gas fields; another twenty-five percent comes from burning of forest trees in the tropics, and the remainder originates largely from rice paddies, domestic cattle and other ruminants, and possibly a small amount from termites.

Craig and his associates also showed that small quantities of methane and hydrogen are being emitted from the deep sea vents on the Mid-Ocean Ridges. Clearly these gases are produced by non-biological processes well below the ocean floor, but we do not know whether they are native to the rising basalt or are being created by chemical reactions with seawater circulating in the ridge-crest system. Craig has estimated that one hundred and sixty million cubic meters of methane are being produced annually from the worldwide ocean ridge crest system, equivalent to about 150,000 tons of oil per year. This relatively trivial output suggests that the ridge crests are not a very promising source of fossil fuels!

Craig has also found helium-3 in the ocean in relatively high proportions to helium-4. Helium-4 is produced in the earth's crust and mantle as alpha particles in the radioactive decay of uranium, thorium, radium, and their intermediate products. Helium-3 is the original isotope of helium which was accumulated during the formation of the earth some four and a half billion years ago. Because it is so light, helium continually escapes from the earth into outer space, and there is an equilibrium in the atmosphere between the rate of escape and the rate of formation of helium-4 from radioactive decay. But if helium-3 were not flowing into the ocean from deep within the earth's interior there should be virtually no helium-3 in the ocean or the atmosphere, because it should have all escaped to outer space. Craig took water samples from the second deepest place in the ocean—the Tonga Kermadec

Trench—and found, much to his surprise, relatively large amounts of helium-3, together with helium-4. The highest ratio of helium-3 to helium-4 was at a depth of about 2,500 meters. He knew that this was the same depth as the top of the East Pacific Rise, and he deduced that the helium-3 was coming in a jet of water from the top of the Rise westward thousands of miles to the Trench. Following the jet eastward across the Pacific, Craig and his colleagues came across an active hydrothermal vent at the top of the ridge, which was spewing out carbon dioxide, hydrogen sulfide, methane, helium-3, and other gases. Subsequent investigation showed that a typical hydrothermal vent community of tube worms, large clams, and sulfide oxidizing, carbon-fixing bacteria lived around the vent. Other similar communities were found in the North Pacific by tracing helium-3-containing jets of seawater back to their origin on the Mid-Ocean Ridge. It has been possible from measurements of the rate of helium-3 escape to estimate the present rate of degassing of the earth's core and mantle. This turns out to be surprisingly high.

Over the past forty years, one of Scripps' senior marine chemists, Ed Goldberg, has concentrated on measuring the quantities of elements that are present in the sea in incredibly dilute solutions, parts per billion and parts per trillion (one part per trillion is one gram per million tons, in volume terms, a speck not much bigger than a little fingernail dissolved in a volume of water equal in size to a block-square forty-storey building). By developing and applying some of the new methods of chemical analysis, he and his colleague, Min Koide, have been able to show that such unlikely elements as platinum, iridium, ruthenium, rhenium, gold, silver, and molybdenum are present in seawater and in marine sediments and organisms. They found that platinum and gold exist in seawater only in parts per trillion instead of parts per billion as had previously been claimed by the great German chemist, Fritz Haber. After World War I, Haber thought he had found enough oceanic gold (in sufficiently high concentrations to be economically recoverable) to enable Germany to pay its post-war reparations. Goldberg and Koide discovered that Haber's gold was largely due to contamination, and that he had been wrong by at least a factor of a thousand. Hence the reparations were not paid—and Adolf Hitler emerged.

During the last twenty years Dr. Goldberg has concentrated on the problems of ocean pollution caused by human activities. In 1975 he proposed a program he called the "Mussel Watch" as a promising method for monitoring ocean pollution. Mussels, like many other lamellibranch mollusks are "filter feeders," that is, they sit quietly in the sea attached to a rock or a pier, and filter seawater through fine screens in their guts. Particles in the water are captured on the screens, and the nutritious ones form the diet of the mussels, the remainder presumably being discarded. In the process, the animals accumulate many different pollutants from the water, and if they are then ground up and analyzed, the quantity and concentration of these substances can be determined, provided the rate of filtration is known or can be estimated. Early versions of the mussel watch were carried out in the United States by the National Ocean and Atmospheric Administration and it has been used by China, India, the U.S.S.R., and many developing countries to measure the degree of pollution of their coastal waters.

Physical Oceanography

Compared to the earth's rivers, the ocean currents transport very large quantities of water. Oceanographers measure ocean water transport in millions of tons per second. This unit is called a "sverdrup" after the great Scripps Director. All the rivers on earth combined, including even the mighty Amazon, carry about one sverdrup, while the Gulf Stream in the Western Atlantic and the Kuroshio off Japan each carry more than seventy sverdrups. These great currents, and others like them throughout the upper layers of the oceans, are part of the wind-driven ocean circulation; their ultimate driving forces are the steady components of our planet's winds—the trade winds of low latitudes and the prevailing westerlies to the north and south. In the North Pacific and North Atlantic the winds exert a drag or stress on the sea surface in such a way that the waters are piled up in huge mounds in the middle of the oceans. Under the influence of the earth's rotation the waters flow around these mounds, parallel to the contours, in great gyres.

Harald Sverdrup was the first to elucidate the dynamics of these gyres. He showed that the gradient of the wind stress with latitude must be balanced by the latitudinal gradient of the "Coriolis Force" due to the earth's rotation, times the north-south component of water transport. His ideas were refined by Henry Stommel of the Woods Hole Oceanographic Institution, who showed that on the western side of a gyre the current must become narrow and the speed must increase, and by Walter Munk of Scripps, who combined Sverdrup's and Stommel's ideas to provide an integrated picture of the wind driven circulation.

The "Coriolis Force" becomes smaller with decreasing latitude and goes to zero at the Equator, hence the stress of the trade winds must be balanced mainly by internal friction in the ocean waters. The result is that ocean currents near the equator are comparatively strong and flow mostly in an East-West or West-East direction. In the tropical Pacific there are five such currents. From North to South they are: the North Equatorial Current, the North Equatorial Countercurrent, the South Equatorial Current, the Equatorial Undercurrent, and the South Equatorial Countercurrent. Three of these currents have long been known from ship reports, but two were found only recently. The Equatorial Undercurrent was discovered in the late 1950s by Townsend Cromwell, a former Scripps graduate student, and the Southern Equatorial Current was later found by Scripps' Joseph Reid. A similar eastward flow south of the Equator was discovered in the Atlantic Ocean on the Scripps "Lusiad" expedition.

John Knauss, then a graduate student at Scripps, made the first measurements of the Equatorial Undercurrent in the Pacific Ocean. He used arrays of current meters, which revealed a strong eastward flow beneath the sea surface, just opposite to a surface flow to the west. Speeds of more than 125 cm/s were found at depths of 100 meters, centered on the Equator. The current is about 300 kilometers wide and about 200 meters thick and carries 40 sverdrups. Later measurements have shown that the undercurrent extends all along the Equator from the Solomon Islands to Ecuador. Bruce Taft and John Knauss investigated whether a similar current exists in the Indian Ocean. They found a much weaker current at the Equator with speeds of 30 to 80 cm/s carrying about 11 sverdrups eastward. This current is present during the northeast monsoon, but it is weak or absent during the southwest monsoon.

Except for visible light, the oceans do not transmit electromagnetic radiation over significant distances; in contrast, sound travels very well in the oceans, a loud, low-frequency sound made at an appropriate depth underwater can be heard for distances of several thousand miles. Consequently, underwater acoustics is an important field of research in many oceanographic institutions. One of the principal laboratories of the Scripps Institution, the Marine Physical Laboratory, has concentrated on this research ever since World War II.

Walter Munk and his partners at Scripps and Woods Hole turned the tables on this research; they are using underwater sound to study ocean currents and the physical properties of ocean water masses. For this purpose they have developed a powerful new technique they call "acoustic tomography." This involves planting a series of buoys containing acoustic transducers (instruments that can both send and receive underwater sound) at a depth of several hundred to a thousand meters around the ocean area they wish to study. The area can cover up to a million square kilometers. By measuring the travel times of signals sent by each transducer and received by all the others, they can determine the average sound velocities along the crisscrossing paths between the buoys. Sound velocity in stationary seawater is principally a function of water temperature and hydrostatic pressure. Hence the measured sound velocities can be translated into temperatures, and by measuring the differences in travel times when the sending and receiving transducers are reversed the average current velocities along the sound paths can be determined. The sound from each transducer is received at each receiver as a multiple signal produced by a series of "sound rays" that have traveled through different parts of the vertical plane between the sound source and the receiver. Thus a three-dimensional, time-varying map of the distribution of temperature and current velocity in the area surrounded by the signal-sending and signal-receiving instruments can be constructed.

The latest use of this new technology has been in the Greenland Sea, where an array of six sound-signaling and receiving buoys was planted in the summer of 1988 and recovered in August 1989. The experiment had a specific objective—to observe and measure the wintertime convective overturning of the Greenland Sea waters, which forms the North Atlantic deep water. The process is hard to observe from a surface ship, as the Greenland Sea is covered with ice during the wintertime.

The technical difficulties of this operation were formidable. To a considerable extent, the usual "perils and hazards of the sea" could be overcome because the buoys were planted at depths well below the zone of wave action. But the instruments had to be entirely self-contained because no contact was possible with the underwater sound equipment for twelve months. An atomic clock, accurate to one millisecond per year was needed to time the acoustic signals. An instrument to produce a sufficiently loud and complex sound signal that could be timed on arrival to a millisecond had to be developed. Electric storage batteries were needed, capable of generating sufficient energy to make a 220 decibel signal every four hours for a year. Finally, each transducer had to be capable of continu-

A new, low-frequency sound transmitter for acoustic tomography is about to be tested. Photo: Peter Worcester

Satellite tracks of twenty-one drifting buoys submerged eleven meters below the surface in the California Current System in 1985 and 1986, for average times of about nine months. Note the intense mesoscale activity of the subsurface circulation. Map prepared by Pierre-Marie Poulain from data of Scripps Satellite Facility

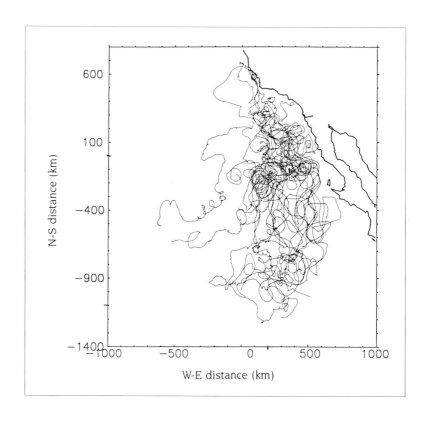

ously locating itself with respect to fixed sound sources planted on the seafloor.

At the time of writing this article the scientific results of the Greenland Sea experiment had not yet been determined: perhaps a year's work will be required to analyze the recording tapes from each transducer. But we do know that most of the buoys worked most of the time, because their signals could be heard from other underwater sound installations.

Oceanographers deal with such vast areas, often under such difficult working conditions, that much of their research is done cooperatively. One recent example of cooperation was an experiment conducted in the Strait of Gibraltar by research teams from Canada, France, the United States, Spain, and Morocco. The Strait is nearly sixty kilometers long and twelve kilometers wide at its narrowest section. It varies in depth from 900 meters to a sill with a maximum depth of 300 meters. Near the surface the net flow is from the Atlantic into the Mediterranean, while beneath the surface, heavy Mediterranean water flows out into the Atlantic. The Scripps team focused on current flows through the Strait, with periods from a few days to a few months. They studied the transport of heat and salt as well as water, the effect of tidal action, and the distribution of water masses (each water mass has unique temperature and salinity characteristics) involved in the exchange between the Mediterranean Sea and the Atlantic Ocean.

Current meters and instruments for measuring temperature, conductivity, and pressure were maintained on moorings along the sill for one year. Water surveys were also conducted from ships. Despite the fact that the Mediterranean is often called the tideless sea, the currents through the Strait were found to be mainly controlled by tidal action, and there are major differences between spring and neap tides. The other principal factors determining the current flow are variations in atmospheric pressure, which are fairly homogeneous over the entire Mediterranean Sea.

An earlier and much larger experiment in international oceanographic cooperation was organized by J. L. Reid of Scripps in 1955 and 1956. This was the NORPAC Experiment; in the summer of 1955 it involved nineteen vessels from the United States, Canada, and Japan, covering the North Pacific between 20° N and 59° N; in 1956 the area between 20 N and 20 S was added, using United States, Japanese, and French research ships. In terms of area covered, these were the largest quasi-synoptic studies ever made; observations extended to about 1,200 meters below the surface. The data they provided led to important findings in the wind-driven circulation, including the existence of poleward countercurrents along the eastern boundaries and Reid's discovery (already referred to) of the eastward flowing South Equatorial countercurrent near 8 to 10° S.

Beside international and inter-institutional collaboration, the nature of the oceanographer's instruments emphasizes the need for interdisciplinary cooperation. The principal instruments are ships, and these are so expensive to build and operate that they must be used as intensively as possible. This usually means that several kinds of scientists must work together to fully occupy an oceanographic ship when it goes to sea.

Both interdisciplinary and inter-institutional cooperation have characterized the longest continuous research program in the history of the Scripps Institution. This is the California Cooperative Fisheries Investigations which was started in 1949 and is still continuing. It involved cooperation between scientists and ships of the United States Fish and Wildlife Service, the California Division of Fish and Game, the Hopkins Marine Station of Stanford University, and the Scripps Institution of Oceanography. They joined together to study the seasonal and annual variations in the biology, physics, and chemistry of the currents and water masses and the weather conditions off California and Baja California.

Spatial and temporal variations at widely varying scales were found in both the NORPAC experiment and the CALCOFI investigation. But the largest variations occurred two years after NORPAC, during the great El Niño of 1957–58.

El Niño, "the Child," so named because it is most noticeable at Christmastime, has been known on the Peruvian coast for hundreds of years, ever since the first ships anchored at the Port of Callao and their hull paint was blackened by the sulfides produced from decaying marine vegetation during an El Niño. But its effects off California and more generally throughout the Pacific were first intensively studied during 1957–58, based on the departures from "normal" conditions observed during the previous eight years in the CALCOFI investigations and the NORPAC Experiment of 1955 and 1956.

Several interrelated events occur during a major El Niño: atmospheric pressure rises in the Western Pacific and falls in the East. The trade winds and upwelling of cold ocean waters in the Eastern equatorial belt cease, and warm waters from the Central Pacific flood the American coasts. Water temperatures off California rise by several degrees Celsius, and mean sea level rises by twenty to thirty centimeters. Torrential rains flood the Peruvian coast, where "normally" no rain falls at all. Hurricanes occur in Tahiti and Hawaii. Severe droughts and forest fires occur over wide areas in the Southwestern Pacific, and monsoons fail in India. Winter climates over most of North America become anomalous.

Biological events in the oceans are equally striking. Because of the cessation of upwelling, phytoplankton production is greatly reduced and there are mass mortalities of pelagic fishes and sea birds. By the fall of 1957 the coral atoll of Canton Island near the Equator, previously bleak and dry, was lush with the seedlings of tropical trees and vines. This transformation was brought about by the unprecedented visitation, earlier in the year, of great rafts of sea-borne seeds and heavy rains.

Since 1957–58 there have been two more extreme El Niños, in 1972–73 and in 1982–83, and smaller occurrences at intervals of about four years. Thanks to the pioneering work of the late J. Bjerknes at Scripps and U.C.L.A., and of many later investigators, the phenomenon is now recognized as a fundamental global climate event which depends upon the interactions between the ocean and the atmosphere in the Eastern Tropical Pacific.

Another expected global climatic event is that anticipated from the Greenhouse Effect—the increasing atmospheric concentrations of carbon dioxide, methane, and other radiatively active gases. The principal uncertainty in estimating the climatic changes results from uncertainties in the thickness, height, and areal extent of clouds as atmospheric temperatures increase. Clouds reflect sunlight, thereby reducing the incoming heat from the sun and tending to counteract the greenhouse effect. At the same time they absorb and reradiate infrared radiation, thereby increasing greenhouse heating at low altitudes.

Using earth radiation data from the Nimbus satellites, Richard Somerville and his associates in Scripps' Climate Research Group have been able to compare variations in the earth's reflectivity or albedo, presumably due to variations in cloudiness, with seasonal variations in sea surface temperature. They have also constructed a theoretical model of the increase of optical thickness in cirrus clouds with increasing atmospheric absolute humidity (which will presumably accompany rising temperatures). This greater optical thickness should enhance the Greenhouse Effect, but they conclude that its effect will be less important than the increased reflectivity and hence cooling, resulting from more cloudiness.

Is world sea level rising or falling? Dr. Tim P. Barnett of the Scripps Climate Research Group has attempted to answer this question by statistical analyses in six different regions of the oceans, where sea level data recorded by tide gauges appears to be relatively homogeneous. His conclusion is that at the present time average sea level in the world ocean may be rising by a few millimeters per year. The six different regions behave differently. South of New England in the Atlantic, sea level is rising steadily at nearly half a centimeter per year; in the Western Pacific sea level appears to have slowly fallen since 1950. This behavior is in marked contrast to that of fifteen thousand to six thousand years ago, when sea level rose on the average by more than one centimeter per year, or to the expected behavior in the future when greenhouse warming will swell the ocean waters, and glacial melting will increase their mass.

Experimental, long-range weather forecasting for the co-terminous United States, on the basis of North Pacific Ocean sea surface temperatures, has long been attempted at the Scripps Institution—during the past twenty years by Dr. Jerome Namias and his collaborators. They have achieved a moderate level of "skill" for forecasts during the fall and winter months. But, more important, they are attempting to clarify the persistence and evolution of atmospheric circulation at time scales of months and longer.

Some of the practical consequences of studies of ocean currents and waves are found in the shallow waters of beaches, estuaries, and harbors. These areas are the focus of investigations by members of Scripps' Center for Coastal Studies, led by Professors Douglas Inman and Clinton Winant. They have studied the origin, transportation, and fate of sands along California beaches. Formerly, these sands were carried to the sea by Southern California's intermittent streams, moved along shore by waves and currents, and lost by slumping, sliding, and density currents down submarine canyons. With the damming and channelization of many of the streams, the sand sources are much diminished, and the beaches are slowly disappearing. One way to avoid this problem is to develop methods for preventing sand accumulation in reservoirs behind dams, and this is being studied by the Scripps group.

In other investigations this group is studying breakwater and seawall structures that dissipate wave energy in the most effective way, using key elements of wave dissipating structures found in nature—coral reefs, grooved bedrock, and natural cliff profiles. They are also developing methods to avoid siltation in harbors around moored or anchored ships.

Scripps research vessel *New
Horizon* coming into San Diego
Bay. Photo: © Lawrence D.
Ford

Geology and Geophysics

In the late 1950s, the late Harry Hess of Princeton University and Walter Munk of the Scripps Institution of Oceanography sat late one night, talking about the need for a dramatic research project in the earth sciences, to match some of the exciting work that was then starting in space exploration and had long been carried out in high energy physics. They conceived the idea of drilling through the deep seafloor far from land, to sample the mantle rocks below the Mohorovicic Discontinuity that separates the earth's crust and mantle. Scripps' Russell Raitt had found previously from his seismic refraction measurements on the Mid-Pac, Capricorn, and later expeditions, that this discontinuity is only a few kilometers deep under most ocean areas, whereas it is thirty-five or forty kilometers deep under the continents. No one has ever been able to drill to the Mohorovicic Discontinuity under a continent, among other reasons because of the high temperature—some 700 degrees Celsius—of the rocks at that depth. But Hess and Munk estimated that it would be possible to drill seven or eight kilometers into the seafloor, where the estimated temperature is only about 150 degrees.

This "Moho" project was taken up by the National Academy of Sciences and the National Science Foundation; a former Scripps staff member, Willard Bascom, was given the task of developing the technology. He and the engineers he recruited did this with spectacular success, demonstrating, in a test several hundred miles west of Guadalupe Island off Mexico, that a drilling ship in four thousand meters of water need not be anchored but could be maintained by underwater acoustic positioning in a fixed position relative to a point on the seafloor, and that a hole could be drilled in the bottom of the sea with ordinary rotary oil-drilling equipment installed on the drilling ship, several thousand meters below the sea surface and several thousand feet into the underlying sediments and solid rock.

The Moho project started off bravely after this demonstration, but it soon bogged down in rapidly escalating costs and congressional in-fighting. After a few years, a new and more modest task was substituted. This was the Deep Sea Drilling Project, designed to drill into the bottom sediments down to the underlying rock throughout the deep oceans and to take core samples of these materials. It was supported by the National Science Foundation, and was coordinated and planned scientifically under a new organization, the Joint Oceanographic Institutions, Inc., consisting of nine United States oceanographic institutions. The Scripps Institution of Oceanography was assigned to manage the project, which began in 1966. By 1980, holes had been sunk at 460 locations throughout the oceans, to depths below the seafloor as great as 5,700 feet, a little over a mile. Thirty miles of cores had been recovered from a total length of holes drilled of nearly 100 miles. Four hundred and eighty scientists from the United States and several other countries had participated in the work at sea, and many others had contributed to the *Initial Reports* of the project. In all, at least two thousand scientist years had been devoted to it.

Among the unexpected results from the core samples was the discovery that massive salt deposits underlie normal oceanic silts and clays in the Mediterranean. These were laid down between five and twelve million years ago. Evidently the entrances to the Mediterranean were blocked during this period and the water in the basin evaporated, forming a huge, mile-deep depression, five times as far below sea level as the present Dead Sea—the deepest spot on earth today—and as large in area as the Indian subcontinent.

Cores from the Atlantic Ocean showed that it did not exist eighty million years ago, and was then gradually formed by the splitting of the great continent of Pangea, first into a series of narrow, stagnant basins in which black shales, salt deposits, and cherts were deposited, and later into the broad ocean we know today.

Several lines of evidence from the cores showed that parts of the ocean lithosphere have moved horizontally for great distances away from the zone of lithosphere formation along the mid-ocean ridges. In the topmost layers of sediments a high concentration of organic remains occurs in a narrow band right along the equator, owing to the high organic productivity of the subsurface waters. In the Pacific this narrow band was found at progressively greater distances north of the equator with increasing depth in the cores, i.e., increasing age of the sediments. These older sediments must have been deposited near the equator, and afterwards moved to their present position. Second, the magnetic dip in older sediments is greater or less than that corresponding to their present position, depending on the North-South component of their horizontal movement since they were deposited. Third, sediments of progressively greater age directly overlie the basaltic basement at increasing distances from the mid-ocean ridges where the basalt was formed.

Estimated velocities of this horizontal motion vary between one and thirteen centimeters per year; some parts of the seafloor appear to have moved

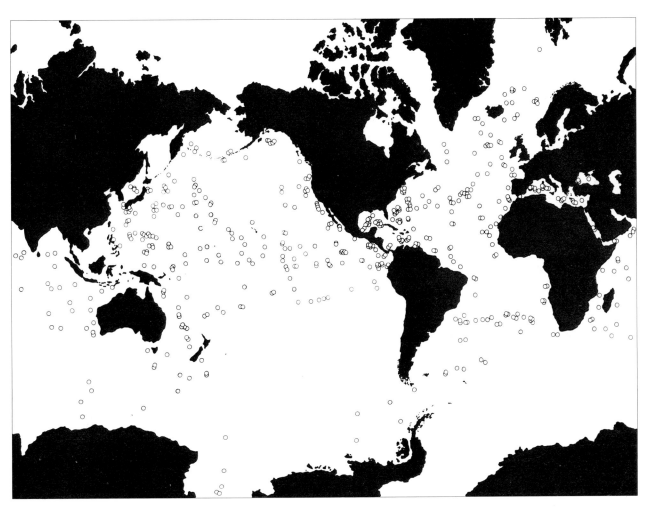

Sites of the Deep Sea Drilling
Project, 1966–83.

several thousand kilometers during the past two hundred million years. As it has moved away from the mid-ocean ridge, the seafloor has subsided relative to the ocean surface at rates up to 60 to 100 meters per million years. This is believed to be due to the cooling and contraction of the sub-oceanic rocks as they moved away from their hot origin along the mid-ocean ridge.

The extreme youth, in geologic terms, of the oceanic lithosphere needs to be emphasized. The oldest rocks and sediments found in any of the 460 boreholes drilled during the entire period of drilling are between 150 and 200 million years old.

In a few cores, sediments laid down at the end of the Cretaceous (the Age of Dinosaurs) and the beginning of the Cenozoic (the Age of Mammals) are found. Large flakes of iron sulfide are found in the cores at this so-called K-T boundary. These are rich in the rare metal, iridium, which in such a locus is believed to be of extraterrestrial origin, strongly supporting Luis and Walter Alvarez' hypothesis that the Age of Dinosaurs ended with the impact of a large asteroid or other celestial body. (In other sediments at the K-T boundary, Scripps' Jeffrey L. Bada has found amino acids of a type found only in some kinds of meteorites, further supporting the impact hypothesis.)

When the Deep Sea Drilling Project began, the drill bit and drill string could not be put back into the same hole once they had been pulled out. Scripps engineers were able to solve this problem of "hole reentry" by a combination of extremely accurate navigation using satellite positioning to locate the drilling ship, and a sonar transponder attached to a large funnel-like cone at the top of the drill-hole. The drill string and bit can now be removed from a hole after it has been well started, and the ship can then sail away to a distant port for repairs. Even after several months the ship can return to within a few hundred feet of the hole, locate the bottom-mounted sonar transponder, reenter the hole, and continue drilling.

The Deep Sea Drilling Project is being managed today by Texas A & M University, using a larger drilling ship and a longer drill string to penetrate the abyssal depths as well as the thick sediments of the Continental Slope. New discoveries continue to be made at a rapid pace.

Although the Project has not yet reached or penetrated to the earth's mantle through the Mohorovicic Discontinuity, upper mantle rocks have been obtained in other ways. They have been dredged from the Tonga Trench and from deep sediment-free rifts in the Mid-Ocean Ridge system of the Western Indian Ocean. This difficult dredging was accomplished by R.L. Fisher and Celeste Engel on a series of Scripps expeditions, starting in 1960 during the International Indian Ocean Expedition, and continuing until the 1980s. Acoustic transponders and underwater photography were used to locate the dredge in relation to the bottom topography. The Indian Mid-Ocean Ridge is particularly susceptible to sampling for deep seated rocks because it is offset by major "fracture zones" or gashes in which the deeper parts of the sub-oceanic rocks are exposed.

Among the dredge samples were coarse-grained rocks that were relatively very low in silica, alumina, sodium, and potassium, and relatively high in magnesium, iron, chromium, and nickel. Their coarse grain showed that they were formed at great depths beneath the seafloor, and their chemical composition almost certainly represents the composition of the earth's upper mantle beneath the Mohorovicic Discontinuity. Similar "ultra-mafic" rocks were later dredged by Soviet scientists from the northern part of the Indian Mid-Ocean Ridge, and highly altered rocks of analogous chemical composition have been dredged from the Mid-Atlantic Ridge.

Four much larger and longer fracture zones than those in the Indian Ocean were found in the Northeastern Pacific in a series of Scripps Institution expeditions during the early 1950s, some of which were conducted in cooperation with a ship of the United States Navy Electronics Laboratory in San Diego. These were described by H. W. Menard. From North to South he called them the Mendocino, Murray, Clarion, and Clipperton fracture zones. They range from 1,400 to 3,300 miles in length and average sixty miles in width. All four are roughly parallel and follow great circles for most of their lengths. The Channel Islands and transverse ranges of California form a landward continuation of the Murray fracture zone and the Revilla Gigedo Islands and volcanic province of southern Mexico form a continental extension of the Clarion fracture zone. These zones are characterized by huge sea mounts, deep, narrow troughs, asymmetrical ridges, and escarpments, two of which are a mile high and more than 1,000 miles long. Menard concluded that deep convection currents within the earth's mantle, caused by differences in heating and cooling, had dragged and stressed the crust, causing it to buckle in narrow bands of plastic deformation, which became the fracture zones. Later work, after the advent of the theories of plate tectonics, has shown that the fracture zones are in part "trans-current" faults related to horizontal displacements of parts of the Mid-Ocean Ridge system.

Scientific progress in oceanography has always

Underwater sea life off the
California Channel Islands: a
Garibaldi fish (*Hysypops
rubicundus*), red gorgonia
(*Lophogorgia chilensis*), and giant
kelp (*Macrocystis pyrifera*).
Photo: © Wayne and Karen
Brown

depended in large degree on development of new technologies for studying the ocean realm. We have described some of these in our discussion of acoustic tomography and of the Deep Sea Drilling Project. Another example is the work of Gustaf Arrhenius and his students and associates on the crystal structures and chemical composition of microscopic components of deep sea sediments that could not be studied with older techniques. This work showed that biological productivity in the equatorial zone of the Pacific was much higher during glacial times than in present and past interglacial periods. Cores ten to fifteen meters long in the top layers of sediments in this zone show a remarkable biogeochemical stratification of calcium carbonate, silica, and phosphatic skeletal remains of marine organisms. The highest levels of nickel, copper, and cobalt from anywhere in the ocean are found in manganese nodules from the northern boundary of this biologically productive equatorial zone, where the skeletal remains are mainly siliceous. Clearly, biological processes are of primary importance in concentrating these valuable metals in the deep sea deposits.

The nature of the earth's interior can be deduced only from signals received near the earth's surface. Two kinds of signals come from earthquakes—direct seismic waves and "normal mode" oscillations of the entire earth. The latter are like the ringing of a complicated bell which vibrates in several different ways. John Orcutt and his collaborators and graduate students have analyzed seismic waves to study the structure of the East Pacific Rise, part of the world-girdling Mid-Ocean Ridge. They find a nearly continuous body of molten magma underlying the Ridge. Freeman Gilbert, Orcutt, and their associates have utilized the normal mode "ringing" of the earth, together with direct seismic waves from earthquakes to obtain a three-dimensional picture of the earth's mantle.

George Backus has used a different kind of signal from the interior—the changing magnetic field on and above the earth's surface, which is caused by fluid motions in the earth's molten outer core. Unfortunately, the surface magnetic fields allow for an infinite number of solutions of the core motions causing the fields. Backus has been able to use measurements of heat flow from the earth's interior into the ocean and other geophysical data to constrain the possible solutions within a usefully narrow range.

Scripps seismologists are contributing to one aspect of the problem of arms control—the detection of underground nuclear weapons tests. Under the joint sponsorship of the Natural Resources Defense Council

(a private American environmental organization) and the Soviet Academy of Sciences, Jonathan Berger and his associates have established three seismic stations each using two different types of seismometers within two hundred kilometers of the Soviet nuclear weapons test site in Kazakhstan. Three other stations are being operated by Soviet scientists within two hundred kilometers of the American nuclear test site in Nevada. Other stations are being installed at much greater distances in the U.S.S.R. to study wave propagation over longer ranges.

A Final Word

During its lifetime, the Scripps Institution of Oceanography has gradually been transformed from a small marine biological station concerned with the marine plants and animals in California's coastal waters to a large earth sciences laboratory concerned with the entire earth and its living creatures. Several external factors have been enormously helpful in this transformation: the University and the State of California, the Office of Naval Research, and the National Science Foundation.

Because it has been for many years an integral part of the University of California, the Institution has been obliged to conform to the University's academic standards in making appointments both to its faculty and to its research staff. At the same time, generous support from the State of California and the Regents of the University has enabled it to expand its faculty to nearly a hundred persons.

More than two-thirds of the Institution's financial support, (some 45 million dollars in 1988 out of a total of 65.7 million dollars) comes from the Federal Government, mainly from ten agencies. The two chief donors are the National Science Foundation (nearly 20 million dollars in 1988) and the Office of Naval Research of the Navy Department (16.7 million dollars in 1988). One of the essential roles played by the National Science Foundation is its support of oceanographic research ships. These vessels are extremely difficult to budget because their numbers of days at sea and the costs of repairs and refitting during any one year are both variable and unpredictable. To reduce this difficulty, the Foundation provides a pool of money for ship support which can be allocated and re-allocated as the needs arise.

From the beginning of Federal support in 1945, until very recently, the Office of Naval Research made an annual grant to the Institution, rather than grants

Scripps research vessel *Melville*
in 1988. Photo: David Wilmot,
SIO

for individual projects. This institutional support was invaluable because it allowed for flexibility and on-the-spot decision-making in the often unpredictable course of research.

In spite of the breadth and depths of its interests, the Institution, in my opinion, is not broad enough. Though its staff deals with many problems of great concern to human beings, the human dimension is pretty much left out of the picture. A greater effort in both research and teaching is needed on the ways in which human beings and their institutions interact with the oceans and their living inhabitants. These interactions are the basis of government policies concerning the ocean realm at all levels of government, from the United Nations to local municipalities. To become a more useful institution to the larger society, Scripps needs to add to its staff some faculty and research staff members who will be able to develop a deeper understanding of how ocean policies can be formulated, implemented, and changed.

2.

The Woods Hole Oceanographic Institution, Massachusetts

Arthur G. Gaines, Jr.

The marine community at Woods Hole is situated on the only natural, deepwater, ice-free harbor on Cape Cod. Starting more than a century ago, the seaside village of Woods Hole has become home to various marine research institutions, as well as to the United States Lighthouse Service and Coast Guard. Photo: A.G. Gaines, WHOI

The Woods Hole Oceanographic Institution (WHOI) is the largest private institution in the world dedicated to ocean research. The Institution has operated up to five ocean research vessels simultaneously, as well as a deep sea research submarine, and five research aircraft. It currently employs more than 900 people, with an annual budget of about seventy million dollars.

WHOI's long-term institutional goal is a humanitarian one: to improve the welfare of humankind through wise use of the oceans. For sixty years the Institution has played a major role in providing an understanding of how the oceans work, with over 7,000 papers published on all aspects of the sea from tropics to poles, and from coastal ponds and marshes to the deep sea, embracing the earth and natural sciences. Beginning in 1971, the social sciences, law, and resource management have been addressed as well, through a center established for marine policy research.

The Oceanographic Institution is international in its scope of research, in the composition of its staff, and in the realm of its collaboration. As a private corporation, the Institution holds a degree of independence not shared by government laboratories; and with research as its highest priority, the staff can sustain a focus and pace uncommon in academia. As a nonprofit organization, the Woods Hole Oceanographic Institution provides a climate where intellectual curiosity drives research, although much of the work conducted here has applied benefits as well. Many of the technologies used to explore the sea and the seafloor, now manufactured and used commercially, were developed at Woods Hole. An education program, operated jointly with the Massachusetts Institute of Technology since 1968, offers graduate degrees in oceanography and ocean engineering. The education program is also geared toward research, with comparatively less emphasis on classroom studies.

Interactions with the mainstream of international ocean science has been an important ingredient and contribution of the Institution. Through informal arrangements possible within its structure the Institution encourages broad collaboration and practical working arrangements for participation in joint research projects. A large number of ocean scientists in the United States and around the world have spent time as visitors or students at Woods Hole. Many of the scientific staff at the Institution have held joint appointments at universities and colleges, and participate in the numerous study committees of the National Academy of Sciences, which for many years has conducted summer programs at Woods Hole.

Staff at WHOI participated in the planning and execution of all of the major national and international oceanographic research programs, from the International Council for the Exploration of the Sea, which started in 1930, to the World Ocean Circulation Experiment, still in its planning stages.

The Woods Hole Village

The economy and the style of life in Woods Hole village have always been intimately linked to the sea. The key natural asset of Woods Hole is a deep and sheltered natural harbor, swept clear of ice and pollutants by tidal currents flowing through an adjacent strait, which also bears the name, "Woods Hole." This navigational passage provides ready access northward to Buzzards Bay and the complex of sounds and bays leading to New Bedford, Providence, and New York in one direction, or Boston (via a canal built in 1914) in the other. Southward, the strait leads to semi-sheltered Nantucket and Vineyard Sounds, used for navigation, fishing, and recreation; beyond the islands of Nantucket and Martha's Vineyard lies the open ocean. From colonial beginnings in 1677 through the eighteenth and early nineteenth centuries, fishing, shipbuilding, and whaling activities flourished locally. In 1863, one of the world's first industrial fertilizer manufacturers set up shop in Woods Hole, to blend agricultural fertilizer from marine phosphate rock from North Carolina, fish meal from New England waters, and other chemical feedstocks from around the world. The substitution of commercial fertilizer for manures produced onsite marked a significant change in United States agriculture. Farming became an industry instead of a craft, in which farm managers began to think in terms of buying fertilizer and selling crops. Ships carried the fertilizer feedstocks to Woods Hole, taking advantage of the natural harbor. By 1895, this malodorous industry had failed at Woods Hole and the industrial sheds had been replaced by some of the most pleasant summer homes on Cape Cod, vacationing families attracted by a cool, summer climate, excellent sailing, and clean water.

The United States Light House Service established a base at Little Harbor in the village to service a growing American merchant fleet. This base was once responsible for eight lightships and twenty-four manned lighthouses in southern New England. Today it still serves buoy tenders and search-and-rescue vessels of the United States Coast Guard. Although a ferry service for Martha's Vineyard had operated from Little Harbor in Woods Hole since 1700, with the construction in 1871 of the railroad (discontinued in 1965) and railroad wharf, Woods Hole became the mainland terminal of the Martha's Vineyard & Nantucket Steamship Authority.

Over the past century, however, the name of this small New England village has been most prominently associated with scientific research in the sea. The first institution located here was the United States Fisheries Commission, built in 1884–85 and originally operated closely with the Smithsonian Institution. Now called National Marine Fisheries Service, this laboratory is operated by the U.S. Department of Commerce. In 1888, the Marine Biological Laboratory (MBL) was founded. It is a private laboratory, originally focusing on summer education programs, soon to expand widely to biological research conducted by summer visitors from colleges and universities. Famous for biomedical research using marine organisms, MBL continues to operate mainly as a summer institution, although a few year-round programs operate under MBL auspices, such as Boston University's Marine Program, The Ecosystem Center, and other programs coordinated by prominent individual scientists.

The Woods Hole Oceanographic Institution was founded in 1930 as part of a national-scale effort to revive oceanography in the United States. A key figure in this effort was the Director of MBL at that time, Frank R. Lillie. The growing concentration of marine scientists and facilities at Woods Hole served to attract additional organizations with marine interests. The United States Geological Survey (U.S.G.S.) came to Woods Hole in 1962, beginning as a single project staffed by five scientists, to survey offshore lands on the United States East Coast. Presently headquarters for the Atlantic Coast and Gulf of Mexico Branch, the U.S.G.S. at Woods Hole occupies buildings rented from

WHOI. The Sea Education Association (SEA), yet another marine organization located in Woods Hole, arrived in 1975. SEA operates an undergraduate maritime affairs program, with courses offered ashore and at sea aboard two sailing research vessels, the R/V *Westward* and R/V *Cramer*. Finally, a number of small research groups have set up operation in Woods Hole, the most recent of which is the Woods Hole Research Center, which focuses on global climate change.

The Shaping of Ocean Science in Woods Hole and America

In the 1870s, the Woods Hole area became the focus of interest for scientists exploring the newly emerging marine sciences. They came from the Smithsonian Institute, Harvard College, and other natural history and education organizations. Most of these scientists were old acquaintances; and their individual and collective efforts led to the establishment of Woods Hole as a national and world center for oceanography. The experience of the first institution here (the United States Fisheries Commission) influenced the focus and modus operandi of the second (the Marine Biological Laboratory), which in turn provided the experience to demonstrate a need for a third (the Woods Hole Oceanographic Institution) and clues as to how it should be structured. Accounts of the history of WHOI seldom adequately credit the other institutions here for their role in shaping it. I would contend that the future of these organizations will be as interdependent as their past.

Two men of uncommon stature in the scientific establishment imparted major characteristics still evident at Woods Hole. One was Joseph Henry, a distinguished American physicist from Princeton University (then New Jersey College) who devoted much of his life to national institution building. Henry was famous among scientists for his research on magnetism, although he remains virtually unknown to the public. He was an architect of the United States Lighthouse system as Head of the Lighthouse Board for three decades (he was responsible for founding the Woods Hole lighthouse base). Henry played the central role in structuring the Smithsonian Institution during its first thirty years; and he was President of the newly formed National Academy of Sciences. It was under Joseph Henry's purview and with his solid, working connections in the government and scientific establishment that his close friend and assistant at the Smithsonian, Spencer Fullerton Baird, founded the United States

The waterfront at Woods Hole
and the Woods Hole
Oceanographic Institution. The
cupola-topped Bigelow
building, built in 1930, housed
the Institution's first research
laboratories and offices.
Photo: Vicky Cullen, WHOI

Fisheries Commission at Woods Hole. Ultimately, the broad marine research and education objectives defined for the Commission by Spencer Baird took half a century and three separate institutions to implement. Following Henry's example, Baird perceived government funding and networking of United States government agency resources as the only realistic way to sustain ocean research. It was the failures attributed to this dependency that caused the other institutions to be formed as private corporations, with greater focus on private support.

The second great scientist who shaped oceanography at Woods Hole was Louis Agassiz, whose example brought to Woods Hole rigorous basic research, an international perspective, excellence in field education, and use of private funding. A Swiss geologist and zoologist, Agassiz was an inspired teacher and lecturer. He had noted the similarity between landforms in North America and those in glaciated parts of Europe and concluded, to the shock of the contemporary world, that vast areas of North America had once been covered by glacial ice. The famous Agassiz came to America on tour at age thirty-nine, and ended up accepting a position at Harvard College. Soon, his scientific interests turned to marine biology, specifically embryology of marine animals, which he thought would cast light on the controversial Darwin theory (that he opposed).

Before there were laboratories at Woods Hole, Agassiz conducted field work here and in nearby waters with Baird and others. In December 1872, toward the end of his life, he issued a circular announcing a "Program of a Course of Instruction in Natural History, to be delivered by the Seaside, in Nantucket, during the Summer Months, chiefly designed for Teachers who propose to introduce the Study into their Schools, and for Students preparing to become Teachers." Agassiz' approach to instruction incorporated a faculty of superb qualification, equal-admission opportunity for women, field research (Agassiz' most famous quote is "Study Nature Not Books"), and instruction tailored to the individual. These elements are still hallmarks of education programs offered at Woods Hole, from the graduate courses offered at MBL and WHOI to the summer field classes of the Children's School of Science, a grade school program.

From news coverage about the course, Mr. John Anderson, a merchant of New York, offered Agassiz a private island (Penikese Island) near Woods Hole to serve as a base of operation for his course, as well as a fifty-thousand-dollar endowment to cover operating expenses. This act of magnificent philanthropy was only the first of numerous instances where businessmen have intervened, often at crucial times, to guide and support the operation of the scientific institutions. The lives of these men and women turned by necessity or chance to business, but often they harbored a life-long love for the sea.

Students gathered together with Agassiz in 1873 at the "Anderson School of Natural History" on Penikese Island, the first American seaside laboratory (one year earlier the great international marine biological observatory at Naples had been founded by Anton Dohrn; fourteen years earlier, the first marine biological station in the world was founded at Concarneau, France). Some of Agassiz' students later played seminal roles at Woods Hole; for example, Alpheus Hyatt later helped found the MBL and served as its first President; and C.O. Whitman later became MBL's first Director. Agassiz' son, Alexander, a professor at Harvard College and marine scientist like his father, would later serve as a mentor to Henry Bigelow, who helped organize WHOI and served as its Director for the first decade.

Agassiz died in 1873 and his seaside laboratory was closed soon thereafter. The failure of the Anderson School contained a lesson that would not be forgotten half a century later when WHOI was established in Woods Hole.

Formation of the Fisheries Laboratory

The founding of the United States Fisheries Commission was logistically the accomplishment of Spencer Baird, a naturalist and curator by training. He had been influenced as a teenager by his mentor and eventual friend, J.J. Audubon, and was qualitative and taxonomic in his approach to natural history. In the course of Baird's long career with the Smithsonian Institution, he spent most of his time stocking a national museum of natural history, during which time he became a skillful administrator and one comfortable at "working" the government legislative and budgetary process.

His life had an idyllic quality, with winters spent at the Smithsonian in Washington, D.C., and lengthy summer stints in the field, vacationing with his family and gathering new materials for the museum at sites around the eastern United States. During the summer of 1863, Baird brought his ailing wife Mary and his daughter Lucy to vacation at Woods Hole, one of their first vacations at the seashore. Baird was evidently much taken by the varied and abundant marine life, a zoological area largely unexplored by him and one under-represented in the Smithsonian collection.

At that time, the commercial fishery in the sounds and bays around Woods Hole depended on operation of over thirty fish pounds and traps. This industry provided manpower and equipment for sampling the wild fishery in a way no scientific study could have matched (then or now). Baird must have been fascinated as the commercial landing came ashore on the docks at Woods Hole. This holiday visit to the village set the stage for development of what has become one of the great world centers of marine biology and oceanography.

The latter part of the nineteenth century was marked by growing worldwide interest in marine biology. New species were being discovered and marine biology stations were being formed in Europe. Inactivity in the United States must have been conspicuous to Baird and others at the Smithsonian. To fund a United States program on marine biology, Baird used a strategy different from any he had used before. He packaged his proposed work to Congress as a fisheries management project, with discrete economic and political overtones, in which the basic research and collecting were included as incidentals. In 1871, Baird arranged to be appointed "Commissioner of Fisheries" by Congress in a joint resolution that he had helped author. The Commissioner, responding to a growing alarm in Massachusetts and Rhode island, was charged with looking into a "drastic reduction" in the coastal fishery, a matter the state legislatures had been unable to effectively address.

During the summer of 1871 Baird went to Woods Hole with a team of scientific colleagues, including the familiar names of Louis Agassiz and Alpheus Hyatt, to carry out the mandate. The formal objectives were based in applied research: To establish the facts regarding the alleged decrease in fisheries; if so ascertained, to identify the causes; and to suggest methods for restoration of the fishery. "Incidental objectives" addressed Baird's broad academic interests in collecting, education, and more basic research, topics which by themselves the Congress would not likely have supported: to study the physical character of the seas and natural history of its plant and animal inhabitants, and to make copious collections for the Smithsonian Institution and colleges and universities around the country.

The use of economic arguments to obtain federal funding for fisheries research started government sponsored research on a course fraught with problems. Over the years it would subject researchers' proposals and results to evaluation in narrow economic terms and a short time frame. This approach reflected Baird's sensitivity to a common, fundamental difference in outlook between those paying for research and those conducting it, a difference that persists today. The approach may also have been Baird's only option at the time.

Results of the summer research in 1871 (using buildings and boats of the Lighthouse Service and other nearby agencies) filled the first report of the United States Commission of Fish and Fisheries, now a brittle and dusty tome that sits on a shelf at the National Marine Fisheries Service. "Incidentals research" includes over 450 pages of the report devoted to distribution of benthic invertebrates—hardly of interest to Congress, businessmen, or fishermen. But Baird also gave them the desired answers in a skillful resolution of the problem: Yes, the fisheries were decreasing; the cause was over-fishing by pound nets and fish traps. The solution was to prohibit fishing between Friday night and Monday morning to allow stocks to pass unimpeded along the shore. The answer was appealing, the cause plausible, and the "solution" practical (it also gave fishermen the weekend off).

For the next decade fisheries research by the Commissioner was conducted at temporary stations along the coast, such as: Woods Hole (1872 and 1875); Noank, Connecticut (1876); Gloucester, Massachusetts (1877); Provincetown, Massachusetts (1880); and, Newport, Rhode Island (1881). In the Report of the Commissioner for 1882, Baird wrote, "The acquisition of a sea-going steamer, in the *Fish Hawk*, . . . rendered it expedient to fix upon some point for permanent occupation where the necessary facilities could be obtained for doing the work of the Commission. . . ." Two sites were seriously considered for the station—Woods Hole and Newport (Rhode Island). An enthusiastic welcome from the merchants at Newport, an offer of Navy land there, and well established shipyards must have made that location attractive. But Baird was concerned about pollution from the cities surrounding Narragansett Bay, seasonal brackishness of the water, and the monotonous muddy bottom which during winter storms clouded the water (and contained a less interesting fauna). Woods Hole was chosen for its clean coastal seawater, needed for aquaculture. The harbor, diversity of marine life, its attraction as a summering spot, and the proximity and railroad access to academic centers in Boston must have weighed in the decision as well.

Land was obtained on the shore of Great Harbor, with the help of colleagues and businessmen, such as

Joseph Story Fay who donated three acres, and John M. Forbes, whose descendants still own the Elizabeth Islands opposite Woods Hole and who have supported the scientific community ever since. In 1884 and 1885 the Commission built a boat basin and the original fisheries laboratory. Research at the Fisheries Laboratory aimed to please a Congress bent on economic results. A major program of the new laboratory, lasting fifty years, was culture and husbandry of marine fish species, especially flounder and codfish, propagated for use in replenishing commercial stocks. Specially designed railcars carried the living fish crop to distant coastal sites for release, although the economic effectiveness of the release program has never been demonstrated.

Following the example of western European laboratories, Baird established the first United States public aquarium—to heighten public interest in marine fishes and awareness of the laboratory. Basic research was also conducted at the laboratory by university researchers, again following the European model: initial contributions by The Johns Hopkins University, Princeton University, and Harvard College endowed "research tables" to be used in perpetuity by academic scientists visiting from these institutions. However, university research at the fisheries laboratory at Woods Hole never reached Baird's expectation, and the full fledged education program Baird had in mind was never to be established at this laboratory. The laboratory and its programs were plagued by financial problems and deteriorating facilities, leading to diminished morale among the staff. The structure of federal control and funding was devastating Baird's seaside laboratory. The Commission somehow limped through these dark years, although its descendant, the National Marine Fisheries Service, continues to suffer from related problems.

The Marine Biological Laboratory

The MBL was established in 1888, based on three simple ideas: the foundation of a permanent biological laboratory at the sea coast, the combination of research and instruction in its activities, and the development of a nationwide interest in its use and support. The initiative was spearheaded by the Women's Education Association, with support by the Boston Society of Natural History, Harvard University, Massachusetts Institute of Technology, and Williams College. The first President was Alpheus Hyatt (Director of the Boston Society of Natural History), and the first Director was

C.O. Whitman. Since their summer together at Penikese Island (sixteen years earlier) Hyatt had run a similar marine program at Annasquam, near Gloucester. Whitman had worked at European laboratories with some of the greatest zoologists of all time, and had been assistant at the Museum of Comparative Zoology under Alexander Agassiz at Harvard.

The distinctions between MBL and the Fisheries Laboratory were clearly stated at Whitman's first annual address. He reiterated the nationwide scope and even international context of the organization, and he urged that all biological interests, not only marine, should be welcomed. In contrast with the neighboring Fisheries Laboratory, Whitman promoted basic scientific research, placing applied or economical aspects at a lower priority. Finally, control and governing of the laboratory should reside with professional biologists. He suggested the new institution be guided by a corps of investigators (eventually termed, "members of the corporation") consisting of the Board of Trustees, investigators conducting research at the laboratory, and MBL employees.

As Director of MBL, Whitman was keen to enlist the interest of large numbers of universities in the fledgling laboratory before similar stations became numerous, which he suspected was in the wind. He soon was successful in enlisting over 150 institutions, of which 80 contributed subscriptions for use of facilities at MBL.

In the first few years, Whitman, a professor at the University of Chicago for the most active part of his life, promoted development of MBL as a national center of research in every department of biology. Lands, temporary buildings and boats were acquired. Once again, local land owners and businessmen played a role. Founders of the MBL are said to have approached Mr. Fay, a local benefactor and pious churchgoer, for aid in obtaining land. This was about twenty years following Darwin's publication of the Origin of Species. Mr. Fay said,

"You are biologists?"
"Yes."
"You believe in evolution?"
"Yes."
"Well, I don't. But I believe, gentlemen, that if you study diligently you will see your errors."

During Whitman's tenure as Director—MBL's first fifteen years—a sharp difference of opinion broke out among the Trustees. They were concerned over the rapid rate of growth and diminishing power of the local

Trustees. The Women's Education Association representatives were uneasy over the increasing financial responsibilities of the Board and the loss of influence in affairs of the Laboratory of those who had founded it. The outcome, after an interlude of tense debate, was that nationwide control prevailed. Periods of financial strife made it patently clear that Whitman was a poor businessman. As a result, MBL considered forming a closer affiliation with a limited number of strong organizations to impart greater stability; or separating scientific control from financial control entirely. In the end, however, these ideas were dropped. Three principles emerged strongly: independence from local control, national organization, and administration by professional scientists. The clarification of this structure was one major contribution of the first Director. Another was making a living reality of Baird's dream: to make Woods Hole a place in which colleges, universities, and research institutions should come together in friendly association and rivalry.

The second Director, Frank R. Lillie, (a student of Whitman), brought the dream to fruition through active growth and solidification of the institution. As a result of his efforts, the MBL has played a major role in the development of biology in America, and is abundantly associated with an impressive list of Nobel Laureates in the biomedical sciences. Not much is written about Lillie, and he does not reveal his inner thoughts in his own writings about MBL. But it is clear that he was effective at marshaling private support for MBL (notably from C.R. Crane and the Rockefeller Foundation), that he understood what scientists need in order to work, and that he knew how to help scientists even when they didn't want it. After his retirement as Director in 1925 (following his seventeen year tenure as Director), Lillie stayed on as President of the Corporation until 1942. Lillie went on to accomplish the last part of Baird's dream—establishment of basic research programs in oceanography at Woods Hole.

The Woods Hole Oceanographic Institution

The movement to create an oceanographic institution in Woods Hole was championed by F.R. Lillie. There were other significant players, but it was Lillie who nurtured the idea, found the financing, structured the institution, identified the first Director, and—as the first President of the Corporation—protected the young institution during the early years.

Basic research in oceanography was not possible at the existing institutions in Woods Hole. The Fisheries Laboratory was kept out of basic research by a budgetary leash controlled by Congress which, in the 1920s, continued to view fisheries research in purely short- and economic terms. From the dilapidated state of the laboratory buildings, ships, and research programs it had become clear that the government was not willing to provide the kind of support needed to sustain offshore ocean research. As Director of the neighboring laboratory, Lillie read this message clearly.

The MBL, while ideally organized to serve summer biological research by individual academic visitors, was unsuitable for oceanography. First of all, oceanography involves sciences not significantly represented at MBL, such as geology, physics, meteorology, and engineering. Secondly, oceanography is viewed by some as an area of applied science, a secondary objective of MBL and no doubt an object of scorn by some of its researchers. Oceanography is far more equipment-intensive than biology was at that time, and the need for ocean-going ships would involve expensive support staff and year-round programs. Most of the scientists attracted to MBL savored the interaction with other scientists in the Woods Hole summer setting—oceanography involved long cruises at sea away from friends, family, and Woods Hole. While MBL biologists pursued independent, intellectually based research in the laboratory, oceanographers must perform physical work in teams on programs they only partly defined.

Equally important, to bring oceanography into MBL would have represented a severe institutional shock to a finely tuned organization, already performing magnificently well. To quote Allyn Vine, "There are two ways to destroy an institution: halve the budget or double the budget." There is no doubt that bringing a large oceanography program to MBL would have been opposed and would have generated an acrimonious internal battle.

It was clear to Lillie that a new oceanographic institution must be formed. It was also clear to him that the institution should be in Woods Hole. Despite differences between the two areas of science, biology did represent an important part of oceanography. Some of the shore facilities of the institutions could be shared, such as the library, which MBL already had. Both institutions would benefit. Finally, since Lillie intended to raise the needed funds and to shepherd the new institution through its formative stages and early years—and he didn't intend to leave Woods Hole—the new institution would have to be here.

R/V *Asterias* is one of a small
fleet of coastal boats operated
by WHOI. The forty-six-foot
Asterias serves as a near-shore
collecting vessel. She is shown
here towing a meteorological
buoy to a mooring site off
Woods Hole for tests. Photo:
Dave Hosom, WHOI

Besides a realistic plan for the new institution, Lillie recognized two additional requirements: a source of substantial funding, and endorsement of the new institution by the scientific establishment. The first of these is obvious, but, given the first, many people would not have recognized the second. For example, Louis Agassiz had not recognized it in the formation of his seaside laboratory at Penikese Island. Without the endorsement even of Harvard College, where Agassiz was employed, the Anderson School failed soon after the founder's death despite adequate facilities and endowment, abundant student applications for summer positions, and a willing teaching staff of distinguished scientists. Lillie paid sufficient attention to history so as not to repeat its errors. He went to lengthy effort and considerable delay to obtain broad endorsement for his new institution.

Funding for the Oceanographic Institution was to come entirely from the Rockefeller Foundation, to which Lillie had strong ties; endorsement of the plan came, with greater resistance, from the National Academy of Sciences, to which Lillie also had strong ties. Lillie's contact at the Rockefeller Foundation was Dr. Wickliffe Rose, for two decades an adviser and philanthropic manager for John D. Rockefeller. Rose was a humanitarian, capable of magnificent philanthropy. Formerly a Professor of Philosophy at George Peabody College for Teachers in Tennessee, he devoted much of his life to alleviating the effects of war on human suffering. Initially, he focused directly on public health and nutrition but then became convinced that, in the long term, education and university research would be the instruments for progress in those areas. He arrived at two special objectives for the Rockefeller Foundation: to enhance the scientific basis for agricultural food production, and to strengthen European universities following the devastation of World War I. Rose became convinced (perhaps Lillie was involved) of the importance of fisheries to agriculture and food production, and the importance of scientific research to fisheries science and other areas of applied biology.

During the 1920s, Rose awarded Rockefeller Foundation grants to the Marine Biological Laboratory for construction of buildings, and to the National Academy of Sciences for fellowships in marine biology. At this time, F.R. Lillie was both Director of MBL and Chairman of the National Academy Fellowship Committee in Biology and Fisheries. Lillie introduced Rose to the staff of the MBL and in 1925 Rose met with the Director of the Fisheries Laboratory, at this time known as the United States Bureau of Fisheries. A young Harvard oceanographer by the name of Henry B.

Bigelow was there as well. The meeting must have been magic, or perhaps Rose had known what he wanted beforehand; he invited from them a proposal for a long-term oceanographic program and an estimate of what it would cost. By 1925, Lillie and Rose had reached a decision to found a new oceanographic institution at Woods Hole.

Although Wickliffe Rose was approaching retirement age, and this would be among his last projects, founding of the oceanographic institution would not be realized until approved by the National Academy, an approval that did not come until Rose had, in fact, retired. It was not until 1927 that the idea of expanded ocean research in America was presented to the National Academy. The Academy formed a Committee on Oceanography—chaired by Lillie—to consider the role of the United States in a world-wide program of oceanography. Funding for the Committee's work was provided through Rose, with an informal understanding with Lillie that the limit of the final cost of establishing an east coast oceanographic institution would be three million dollars. The Committee brought in Henry Bigelow to spend a year making investigations on their behalf (later published as a book).

Oceanography, Its Scope, Problems, and Economic Importance

One wonders how often the Rose-Lillie plan was threatened by derailment through committee members who did not share the predilection toward a Woods Hole site, or toward the founding of a new institution to stimulate oceanography.

In 1929, the Committee issued their report, containing the essential ingredients formulated by Rose, Lillie, and Bigelow. The report underlined the importance of pursuing oceanography on both United States ocean coasts. It suggested creating a few new institutions and endowing a few existing ones (Scripps Institution of Oceanography of the University of California, the Hopkins Marine Station of Stanford University, and others) on the west coast, rather than attempting to coordinate a large number of university or government programs. It recommended support of a "truly oceanic" station (the newly formed Bermuda Biological Station) and a sub-arctic oceanographic station.

The report placed highest priority on the Atlantic Oceanographic Institute. The sequence of events leading to formation of the institution at Woods Hole suggests almost frantic pushing and pulling to complete the Committee's work, and Lillie's hand in this is evi-

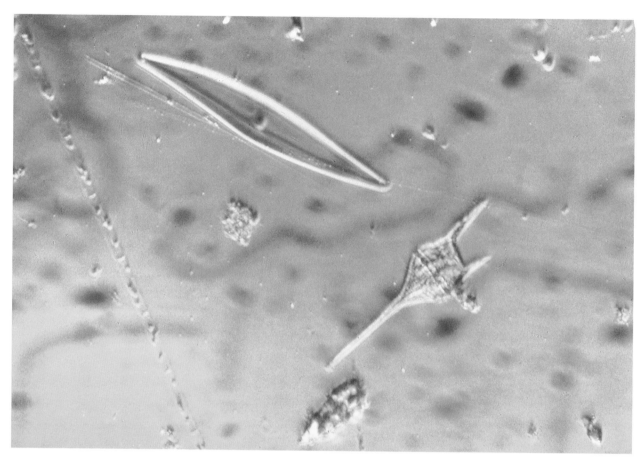

Microscopic plankton algae
represent the primary
producers of the ocean,
ultimately providing virtually
all the food for animal life in
the sea. Plankton may also
release gaseous forms of
sulfur into the atmosphere,
contributing to the
"greenhouse effect" upon
global atmospheric
temperature. Dinoflagellates
related to that shown here
(right of center), can cause
toxic red tides and ciguatera
fish poisoning. Photo: Donald
Anderson, WHOI

dent. A week before the Academy's recommendation was issued, trustees of the Rockefeller Foundation authorized its Executive Committee to aid in the construction and support of an Atlantic Oceanographic Institute, should that recommendation be adopted. Two days after the Academy's report was issued, the Committee on Oceanography agreed the Marine Biological Laboratory at Woods Hole should furnish a site (although other sites had been mentioned). Six weeks later, on January 5, 1930, the Woods Hole Oceanographic Institution was incorporated; and only ten days afterward the Rockefeller Foundation indicated they would be glad to consider a formal request from the Woods Hole Oceanographic Institution for construction and operation funds, a request that was submitted and approved the next month.

While superficially the Woods Hole Oceanographic Institution "sprang full-blown into existence like the goddess Pallas Athena . . ." complete with endorsement, leadership, and funding, it was the culmination of many years' planning and institutional experience. The first Director was Henry Bigelow. The Board of Trustees was comprised of representatives from Harvard University, the United States Coast and Geodetic Survey, Princeton University, the University of Wisconsin, the University of Chicago, the Scripps Institution of Oceanography, the Carnegie Institution of Washington and—to insure fiscal competency—four "men of affairs": Newton Carlton, Seward Prosser, Lawrason Riggs, Jr., and Elihu Root, all of New York City. F.R. Lillie became first President of the Corporation.

Rockefeller's Executive Committee appropriated two million dollars, of which half was for construction, boats, and equipment, and half for an operational endowment. It was hoped this endowment would operate the facilities in perpetuity. An additional $500,000 was to be provided over a ten-year period for current expenses at a rate of $50,000 per year. In 1935, this arrangement was amended by an additional one million dollar endowment, provided in place of the annual appropriation.

Henry Bigelow—First Director (1930–1940)

Henry Bigelow was an oceanographer in America when there were few who chose that appellation. His research on the oceanography of the Gulf of Maine, conducted mostly as a summer investigator from Harvard University aboard deteriorating ships of the United States Bureau of Fisheries, remains a classical

oceanographic model. Although Bigelow was slight in build, he must have been robust to withstand the rigors of life aboard the fishing schooner in the Gulf of Maine, which he described as ". . . a mare incognitum . . . so far as its floating life was concerned."

The Oceanographic Institution's new building and ship were ready in about one year. In the first decade, WHOI's operation emphasized ship-based and summer programs. Bigelow gave the institution research program early momentum by bringing in "ringers." Albert Parr, a Yale oceanographer and licensed seaman; Floyd Soule of the International Ice Patrol; Carl Rossby, an MIT meteorologist; Selman Waksman, a prominent soil microbiologist; Norris Rakestraw, a chemist from Brown University; George Clark and Alfred Redfield from Harvard University. The fledgling staff brought their students along with them: Mary Sears, a student of Bigelow, and Columbus Iselin, a former student; Bostwick Ketchum, a student of Redfield; Athelstan Spilhaus and Raymond Montgomery, students of Rossby; and others.

The headquarters of the Institution were considered to be aboard the vessel R/V Atlantis, and during the winter Bigelow sometimes found himself alone in the building at Woods Hole that now bears his name. Nevertheless, Bigelow's determination propelled programs that addressed most of the major disciplines of ocean research. A New England blue blood, Bigelow provided a solid foundation for the later growth of the Institution. Friends have described him as a Yankee who talked like a New England countryman—"t'aint so"—and a well informed naturalist, well traveled, and a family man.

Bigelow felt oceanographic research must in the end serve economics—in his day to improve fisheries catch and weather prediction. As it has developed over the years, oceanography now serves the ultimate economic issue, to insure the long-term global habitability of the Earth.

During his final illness, Bigelow was visited at the hospital by two of his colleagues from the Oceanographic Institution. Henry eyed them at the bedside and said, in a thin whisper, "Why aren't you fellows at sea?"

Columbus Iselin—War Years (1940–1950; 1956–1958)

A student of Bigelow and first Captain of the R/V Atlantis, Iselin was a physical oceanographer. He led the Institution through two crises, rapid growth during

World War II, and contraction following the war. From employment of 60 personnel in 1940, the Institution grew to 335 people by 1945. Iselin effectively turned the Institution's focus to support the war effort. Studies of sea gull flight behavior became tools for understanding atmospheric turbulence and naval smoke screens; academic research on sound in the sea revealed tools for detecting submarines and signaling underwater; invertebrate biologists turned to antifouling research, credited with saving ten percent of warship fuel consumption; and new programs on underwater explosives were initiated.

A major Institution challenge arising during Iselin's era was increasing dependency upon funding from the Federal Government. During the war, Navy funding for the first time began supporting basic research. Unlike universities with state appropriations, tuition, or large endowment income, WHOI research still depends largely on grants and contracts from the Navy and the National Science Foundation, which together have comprised about eighty percent of the budget for many years. In the end, Joseph Henry appears to have been correct.

Edward H. "Iceberg" Smith— Broadening Vistas (1950–1956)

Iceberg Smith, the third Director, had become a member of WHOI's Trustees in 1933. His nickname "Iceberg" came from a career in the United States Coast Guard in which he played a major role with the United States Ice Patrol. For years Smith had been stationed at the Woods Hole Coast Guard base. Born in Vineyard Haven, on Martha's Vineyard, he was valedictorian of the Class of 1909 at Tisbury High School. He came from a Cape Cod family dating back to the Pilgrims; one ancestor had operated the ferry from Vineyard Haven (then Holmes Hole) to the mainland in 1756. Smith was known for his Arctic voyages aboard *Marion*, and early exploration of the Davis Strait, Baffin Bay and the iceberg-producing glaciers of Greenland. He participated in a voyage to the North Pole aboard the *Graf Zeppelin* in 1931, as navigator, and when over the pole he quipped, "Now head south."

Iceberg Smith was sixty-one years old when he became Director of WHOI. Under his administration, oceanography in America was growing rapidly, as the National Science Foundation expanded its emphasis on basic research in the sea. As the Institution broadened its programs and vistas, with cruises now venturing beyond the Gulf of Maine and the Western North Atlantic, Smith tried to provide increasing institutional structure. He encountered the classical problem of organizing scientists' activities in a growing institution, an effort often viewed by staff members as an infringement on academic freedom. His proposed departmental organization structure was turned down by the Board of Trustees. Smith was known for his Friday afternoon "white-glove" inspections of facilities. Among his successes, Smith organized the WHOI Associates Program, now including about 1,600 members, which has brought both friends and private funding to the Institution.

Paul M. Fye—Giant Oceanography (1958–1977)

The nineteen-year directorship of Paul Fye encompassed the period of most rapid sustained growth in the Institution. His years as Director embraced the initiation of large ocean programs, conducted by teams of investigators from institutions around the world, with national facilities and coordinated ship fleets. This period well illustrates the magnitude of uncertainty and challenge facing a director. Before accepting the position, Fye informed the Trustees he had no experience or interest in raising money, an activity they agreed was neither needed nor appropriate for the Director. In the end it was as much his skill at fund-raising as any other that characterized Fye's contribution to the Institution, with an eight-fold increase in the endowment and major capital enlargement. A second, unanticipated challenge was the growth in oceanographic graduate education in America, with over 100 universities creating graduate programs or departments of oceanography.

The informal arrangements for visiting students at WHOI, in which the student would conduct research at Woods Hole and obtain their degree from the parent institution, were no longer viable. In 1968, a joint MIT-WHOI graduate program was formed. To accommodate the size of ocean programs, the Institution itself had to grow. Sailing ships of the past were replaced with large motor vessels; shore laboratory space increased elevenfold and addition of the new 180 acre Quissett Campus provided space for the foreseeable future. The growth in facilities, employees, and programs meant that greater organizational structure—departments, a tenure system, administrative responsibility—could no longer be put off. Like Iceberg Smith, Fye encountered stiff resistance and even a palace revolution; but in the end he prevailed.

Researchers are evacuated
from their Arctic ice camp.
During a 1989 field program
the camp drifted perilously
close to open water and had
to be abandoned. Photo: Keith
von der Heydt, WHOI

Mary Sears, now a Scientist
Emeritus and Honorary
Trustee of WHOI, came to the
institution in 1933. An
invertebrate biologist during
the early years of her career,
Sears co-founded and served
as editor of *Deep-Sea Research*, a
prominent scientific journal.
During World War II she
founded and was first Director
of the Naval Oceanographic
Unit, and she was the
organizer of the first
International Oceanographic
Convention in 1959. Dr. Sears
is shown here in 1960. Photo:
Redwood Wright, WHOI

Like directors before him, Fye recognized the need for the Institution to address societal problems. He founded the Marine Policy Center to embrace the study of man's exploitation of the ocean by all nations, and the need "to use the oceans wisely."

Fye initially came to Woods Hole during the war years, riding the ferry from New Bedford. He was responsible for research on submarine explosives and worked with a team including several people (such as James "Spike" Coles) who would become loyal supporters of the Institution. Ruth Fye joined the war effort at WHOI as a secretary, no doubt little expecting she would devote over thirty years of her life to WHOI as an advisor to her husband and ambassador of the Institution.

Although he was a firm administrator who would fight for his ideas, Paul Fye was also a kind person. He described himself as "a worried optimist"; his goal "to help the research staff accomplish their career objectives" was also a mechanism for institutional operation and growth. Fye recognized the commitment and wide variety of skills needed to run a ship-operating institution. He was as comfortable on the docks as he was in the board rooms.

John M. Steele—Dollar Inflation and Government Program Competition (1977–1989)

Inflation in the United States beginning in the 1970s made sustained rapid growth at WHOI increasingly difficult in the 1980s. Responding to the nationwide outcry over pollution of the environment, including the oceans, Congress and the President established numerous Federal environmental programs and agencies and enlarged others. The Environmental Protection Administration (EPA) and the National Oceanic and Atmospheric Administration (NOAA) were two of these. While the growth of interest in the oceans could only be considered good, it meant competition for tax dollars for programs dependent on NSF and the Navy, such as the Woods Hole Oceanographic Institution.

It was in this climate that John Steele assumed the Directorship of WHOI. With his appointment, the Trustees of the Institution reaffirmed their dedication to the basic founding tenets of the Oceanographic Institution, although they had little advice as to how to meet the challenges that lay ahead.

John Steele, currently President of the WHOI Corporation and a Senior Scientist at the Marine Policy Center, is a native of Edinburgh, Scotland, and spent most of his career at the Aberdeen Marine Laboratory, eventually as its Deputy Director. Steele's direct participation in research, applying mathematics to physical and biological processes in the sea, had been more active than that of WHOI Directors before him. In 1973 he was awarded the Alexander Agassiz Medal by the National Academy of Sciences for contributions to theoretical and practical studies of factors controlling basic productivity of the oceans.

John Steele's career has also been marked by an extremely active international quality, with participation in numerous international organizations and international oceanographic experiments. He has been a member of the International Council for the Exploration of the Sea (ICES), the Intergovernmental Oceanographic Commission of UNESCO, and the presidential Arctic Research Commission. His collaboration with Woods Hole scientists began in 1958.

A natural system modeler, Steele is aware of the importance of interactions in the sea. In response to balkanization of WHOI along academic disciplinary lines—biology, chemistry, geology, etc.—he founded centers, such as a Coastal Research Center, to reestablish cross-disciplinary thinking.

Craig Dorman—The Future (1989–)

Craig Dorman retired from the Navy as a Rear Admiral to become the Institution's sixth Director. Dr. Dorman was an early graduate of the WHOI/MIT Joint program, in which he excelled academically, receiving the doctorate in physical oceanography. Included in his background in the Navy was membership in a SEAL (Sea Air Land) unit in Vietnam, a Naval analog of the Green Berets, including air deployment as a diver for underwater demolition, and many years' involvement in antisubmarine warfare programs.

Dorman comes from a background of physical and mental discipline, challenge, and excellence. He is convinced of the importance of the ocean in the global flux and balance of energy and materials, and the importance of WHOI as a valuable national asset. The future holds gigantic challenges to the Institution, as well as the opportunities that inevitably accompany them. Increasing growth of mission agencies such as EPA and NOAA with increased competition for NSF and ONR funding pose known challenges. The implications of reorganization in Europe and the Soviet Union are less clear, except that growing international linkages are likely.

The R/V *Atlantis*, designed by
WHOI for ocean research, was
the Institution's first ocean
going ship. Retired in 1966,
after 799 voyages, the *Atlantis*
is still depicted on the WHOI
logo. Photo: WHOI

The R/V *Knorr* is WHOI's
largest research vessel. Her
ice-strengthened hull allows
the *Knorr* to operate in pack
ice; she is shown here near
the South Sandwich Islands
on an Antarctic cruise in
February 1984. Photo: R.J.
Bowen, WHOI

Research at Woods Hole Oceanographic Institution

Ocean research and engineering is the principal focus of WHOI. Each person no doubt holds his own views concerning the Institution's major contributions over the years. I have been guided in the following selection by views expressed by the late Roger Revelle, for many years the Director of the Scripps Institution of Oceanography.

The Gulf Stream

A logical ocean feature for physical oceanographers at Woods Hole to begin studying was the Gulf Stream, the most prominent ocean current of the North Atlantic, passing a short distance offshore from Woods Hole. The Gulf Stream was first charted by Timothy Folger, a whaling captain from Nantucket, whose chart was published by his cousin, Benjamin Franklin.

Gulf Stream research, begun at Woods Hole by Columbus Iselin, has involved some truly great ocean scientists, such as Fritz Fuglister, Valentine Worthington, and Henry Stommel. These men provided descriptive and theoretical understanding of oceanic gyres and other water movements, applicable to all oceans.

Scientists at the Oceanographic Institution demonstrated that the Gulf Stream is deep and narrow, meandering within the broad zone mapped by Folger between the Sargasso Sea and colder coastal waters. The volume of water that flows in the Gulf Stream, although variable, is about one hundred times that of all rivers on Earth combined. In his 1948 theoretical work, Stommel addressed a particularly striking aspect of this ocean current, its intensified speed along the western margin of the North Atlantic Ocean, along the United States East Coast. From theory, Stommel predicted that this kind of ocean feature should occur on a rotating celestial body where continents obstruct the current flow. He showed that the Gulf Stream was not an oddity associated with the western North Atlantic, but that similar currents should exist in all the world's oceans bounded by continents, regardless of whether the western current flows north or south.

In 1960, Stommel published a theoretical analysis predicting that currents at the deep seabed flowing toward the equator should also become intensified along the western margin of the ocean basin. Currents generated by the sinking of cold, highly saline water at high northern latitudes could generate strong currents

along the continental margin of the United States East Coast at more than a mile depth. Later in the 1960s, Charles Hollister, a geologist and currently Vice President and Associate Director of the Institution, photographed the effects of these currents on bottom sediments—ripple marks—and then directly measured the currents off New England. He showed that, in contrast with the prevailing view of a quiescent deep sea, strong currents could actually erode and transport sediment, building major deposits elsewhere, such as the Blake Spur near the Bahama Islands.

This work has major implications regarding human use of the sea. For example, it indicates that wastes disposed in certain ocean areas will not remain there but instead can be moved about by the deep currents.

The Influence of Marine Life on Seawater Composition

Although plants and animals appear widespread in the sea, their total mass is small compared with the vastness of the ocean. Many scientists assumed, therefore, that the overall influence of organisms on the chemical composition of the sea must be negligible. One of the many significant contributions of Alfred Redfield, an oceanographer, biologist, and Associate Director of WHOI, was to show that the activities of marine life control the concentration of nutrient elements in the sea. Redfield found that the relative proportions of carbon, nitrogen, and phosphorus in planktonic marine life occurs in a fixed proportion of about 106:16:1, known now as the "Redfield Ratio." This ratio was discovered through chemical analysis of marine life captured in fine mesh nets from many sites in the ocean. Redfield found that when marine life dies and sinks, the elements are released through decay and digestion in identical proportion—106:16:1—in the form of dissolved materials in seawater. Since decomposition also normally involves oxygen uptake and carbon dioxide release, these substances also change in a fixed and proportional manner.

Redfield's work in this area has served a practical purpose in our understanding of how decomposition of organic matter can affect the composition of natural waters. The work provides a basis for interpreting certain chemical changes we observe. It relates water chemistry to fundamental ecological principles, articulated by other scientists, such as the idea that the growth of marine life can be constrained or "limited" by inadequacy of a single nutrient element, even if

The Nauset embayment, on Cape Cod, has a dynamic tidal inlet and provides a natural laboratory for studying sediment transport and other beach processes. WHOI scientists have also studied salt marshes and the impact of nutrient enrichment at this site. First mapped by Champlain in 1605, the Nauset embayment has undergone many changes that provide insights into how natural coastal systems work. Photo: A.G. Gaines, WHOI

others are abundant. Conversely, addition of a single nutrient element to natural waters can result in rapid growth of algal material if that nutrient happened to be the one limiting growth, accounting for algal blooms in response to pollution by phosphorus-containing soaps.

Extensions of Redfield's work resulted in more complex models of the chemistry of seawater, such as its ability to neutralize acids, or to the production of hydrogen sulfide or methane in sediment porewater. Starting with the Redfield Ratio, other researchers have found variations related to the growth rate of algae, suggesting a means for assessing the "health" of natural populations.

Temperature in the Ocean

The progress of oceanography strongly depends upon instruments, the extensions of man's limbs and senses that permit him to work in and sample the sea. Prior to the 1930s, temperature measurement in the sea depended on deployment of chains of thermometers made of glass and mercury, which produced accurate but discontinuous records of the temperature profile. This was a slow and laborious process with incomplete results. In 1934, C.G. Rossby, a Professor of Meteorology at Massachusetts Institute of Technology and staff member at WHOI, announced the development of the "oceanograph" or "baro-thermograph" for continuous recording of vertical temperature in the upper 200 meters of the ocean.

The instrument was based on the Jaumnotte meteorograph, deployed by balloon for atmospheric studies. Athelstan Spilhaus, currently an Honorary Trustee of WHOI, redesigned Rossby's frustrating contraption and renamed it the Bathythermograph or BT. A working model was available by 1937. The BT, which produced a continuous profile of temperature versus depth, could be deployed from a rapidly moving ship and lowered to hundreds of feet depth. This early oceanographic instrument became a basic tool for studies of surface layer temperature, mixing, and ocean currents, which could often be traced using the distribution of temperature. The BT rapidly accelerated our ability to collect large volumes of temperature data near the ocean surface. The Woods Hole Oceanographic Institution has published whole atlases of BT records from around the world's oceans.

Maurice Ewing, who worked at WHOI before joining the Lamont-Dougherty Geophysical Laboratory in New York, and Allyn Vine, now Scientist Emeritus at WHOI, made improvements on the BT and designed versions for attachment to the hull of submarines during World War II. Ewing and Iselin showed how BT records could be used to identify areas of thermal stratification, a kind of ocean feature that deflected the sound waves used to search for submarines. Thus a submarine could hide under these areas, safe from detection. The BT became an important tool used aboard submarines to evade SONAR, and thousands were used both on antisubmarine ships and on submarines themselves.

During the mid-1960s the Expendable Bathythermograph (XBT), invented and manufactured for the Navy by Sippican Corporation, largely replaced the BT. These small, inexpensive, metal and plastic torpedoes, which can be deployed over the side of a ship, drop at fixed speed through the water and indicate temperature by the change in pitch of an electronic tone, emitted by a temperature sensor. The signal is carried to the surface through a fine wire that unravels from spools on the sinking torpedo. Deeper and faster temperature profiles became possible with the XBT, but as Spilhaus has said, with some disgust, "They litter the bottom with all that wire, plastic, and lead."

In 1971, Neil Brown, currently a Senior Research Specialist at WHOI, invented a Conductivity-Temperature-Depth measuring device, the CTD, which electronically measured salinity and temperature along continuous profiles to the bottom of the sea. This tool, lowered into the sea on a hydrographic wire or conducting cable, provided oceanographers with an electronic measurement of salinity, which together with temperature and depth allowed oceanographers to determine the density of seawater at any depth. Density is essential information for calculating the driving forces of ocean circulation. The invention of the CTD eliminated the need for oceanographers to collect water samples to study ocean circulation, making obsolete the image of the bare-chested oceanographer attaching a Nansen water bottle to a wire. Brown founded his own company to manufacture and market the CTD, improved versions of which are a basic tool of modern oceanography. The CTD has become a major tool for physical oceanographers. Henry Stommel, Senior Scientist at WHOI, recently received the Presidential Medal for Science for his contributions to understanding the physics of ocean circulation. One of his principal tools for obtaining measurements in the last twenty years was the CTD.

The final chapter in this story is the recent development of a free falling CTD that stores information in an on-board computer. This streamlined device, the

Alfred Redfield, one of the first WHOI staff members, epitomizes the multi-disciplinary oceanographer. Originally an animal physiologist, he is known for his research on the influence of organisms on the chemical composition of seawater; and he also recognized that salt marsh peat preserves information about the rising sea level over the past ten thousand years. Photo: WHOI

Researchers deploy
"Mochness," a set of large
nets mounted on a frame that
can be remotely triggered to
open or close at any desired
depth. Mochness is used to
sample invertebrates and
small fish. The cylinder on the
frame measures temperature,
salinity, and depth, and sends
the information back to the
ship. Photo: Peter Wiebe,
WHOI

During his long career at WHOI (beginning in 1937), Allyn Vine has made major contributions to the design and production of ocean technology. Photo: WHOI

"fast hydrographic profiler," was developed by Albert Bradley, a Research Specialist at WHOI, and his colleagues. It carries a disposable weight to bring it to the ocean bottom where the weight is detached and the profiler returns to the surface, rising by buoyancy forces. In transit, the instrument records salinity, temperature, and depth, storing the information in a small computer. Tail fins activated by acoustic sensors in the nose guide the profiler back to a sound-emitting device deployed alongside the ship. Because of its sometimes dramatic reappearance at the ocean surface, the fast hydrographic profiler is also called "Flying Fish."

Sound in the Sea

Even before hydrophones were available for scientists to place in the ocean, sailors knew the sea was filled with sounds. Anchored in shallow water, they could hear the clicks of shrimp and the grunts and buzzing of fishes through the hull of the ship. Maurice Ewing and his colleagues at Woods Hole did some of the earliest and most rigorous research on sound in the sea early in World War II. From this work, focusing on how sound could be used to detect submarines, it became clear that sound was of great usefulness in studying the sea and seafloor, and for signaling within the sea. While light is rapidly absorbed in ocean water, sound can be conducted for many thousands of miles. The acoustic depth sounder is only one of the better known applications of sound for studying the ocean and for use aboard boats and ships of all sizes.

The Navy initiated certain acoustic research because sound-activated mines deployed in coastal waters against enemy submarines were inexplicably detonating, when neither ships nor submarines were nearby. Bio-acousticians at Woods Hole, such as Bill and Barbara Schevill and their protégé, Bill Watkins (presently a Senior Research Specialist at WHOI), have recorded the energetic sounds produced by marine animals, such as whales, that may have detonated an acoustic mine. These researchers have recorded the characteristic sound "prints" produced by over forty species of marine mammals—clicks, squeals, moans, and grunts—used for signaling and echo location. The exact meaning of these sounds and their patterns is still largely unknown, although it appears animals have long used sound for signaling in the sea.

Discovery of Meso-scale Eddies

The Bathythermograph made it possible to define the margins of the Gulf Stream with greater precision than

ever before. Fritz Fuglister, a physical oceanographer at WHOI, and others, found that meanders in the path of the Gulf Stream, like those in the path of a river, would occasionally break off (similar to oxbow lakes near rivers) and form separate rotating eddies measuring 75 to 150 miles in diameter. Typically, the eddies formed south of the Gulf Stream have a core of colder water, entrapped during the meander process. Using more modern technology, Phil Richardson, a Senior Scientist in physical oceanography, has deployed satellite-tracked surface buoys in the eddy rings and watched daily for months as the positions of the buoys go round and round. Gradually these eddies move southward where they are absorbed back into the Gulf Stream, usually south of Cape Hatteras. Because the speeds of the eddy rings can be quite strong, some sailors believe that winning the biennial Bermuda Race may depend upon encountering a cold-core eddy on the limb that carries them toward Bermuda, rather than that detracting from their progress.

Eddies also form north of the Gulf Stream, in which case they have a warm core of Sargasso Sea water containing associated marine life. Peter Wiebe, now Chairman of the Department of Biology, and others in a major Warm-Core Ring program have studied the biological developments in warm-core eddies, as ocean communities of fishes, invertebrates, and algae trapped within these meso-scale ocean features age, eventually to reenter the Gulf Stream many months later, in this case near Chesapeake Bay.

It is now known, through major international cooperative experiments, that the oceans are filled with meso-scale eddies, which are formed in various ways. The drive to understand and map them has motivated a major effort in multi-vessel cruises, international cooperative research, and deployment of complex arrays of instruments over hundreds of square miles of ocean. One acoustic methodology (acoustic tomography) performs a sort of "CAT SCAN" on the ocean, by simultaneously shooting a large number of sound impulses through an area of ocean between a large array of hydrophones (underwater microphones) and sound-producing transducers (underwater speakers). The complex pattern of delays and arrival times of these sound impulses, processed by computer, reveals the three-dimensional distribution of eddies and ocean "fronts."

A particularly successful and interesting application of sound has been the SOFAR float. Buoys weighted to sink to predetermined depths between 200 and 3,000 meters are equipped with powerful batteries and transducers. As these SOFAR floats drift for

The "fast hydrographic profiler," also known as the "Flying Fish." Photo: Al Bradley, WHOI

Dolphins play at the bow of a research vessel, south of New England. Marine mammal research at WHOI ranges from metabolic, learning, and behavior studies to investigations concerning their use of sound for communication and navigation. Photo: A.E. Brearley

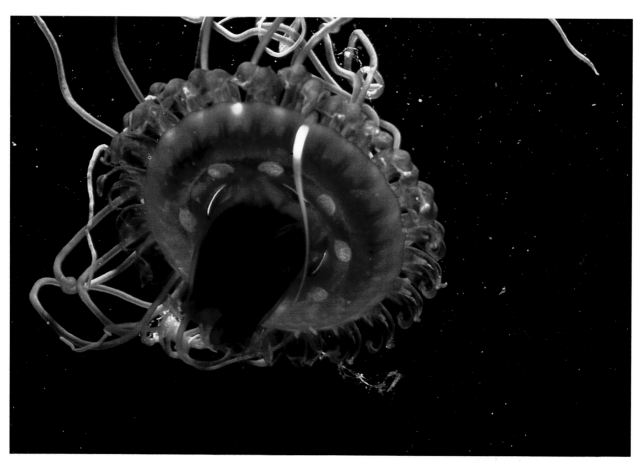

Jellyfish such as this one
(*Atolla wyvillei*) drift at mid-
water depths in the ocean,
worldwide. Its bell-shaped top,
sometimes eight inches in
diameter, moves forward amid
outstretched tentacles, while
the trailing deep-red mouth
awaits the arrival of dinner.
The red shades may be
attributed to pigments in its
diet of marine crustaceans.
Photo: L.P. Madin, WHOI

A small oceanic jellyfish (*Periphylla periphylla*), also found at mid-water depths, illustrates the remarkable variability in color and form of creatures that share very similar habitats and feeding styles. Photo: L.P. Madin, WHOI

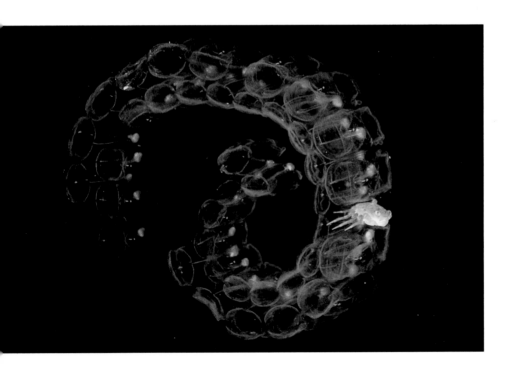

A sargassum crab (*Planes minuta*) clings to a substitute home within a chain of salps (*Pegea socia*). The inch-long crustacean usually occupies clumps of floating sargassum weed, where it finds both food and shelter. The delicate salp, or tunicate, is a colonial filter feeder and requires special sampling techniques to avoid destruction of the chain. They reproduce by budding, one animal pinching off to add another unit to the chain. Photo: L.P. Madin, WHOI

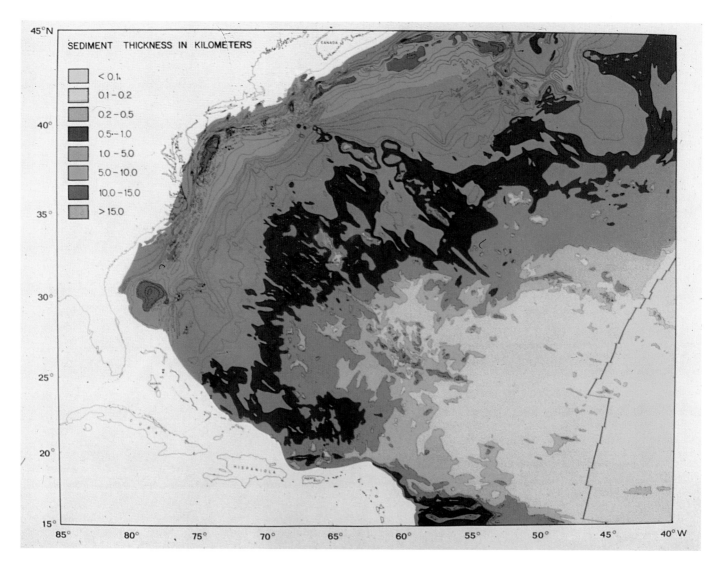

SEDIMENT THICKNESS IN KILOMETERS

- < 0.1
- 0.1 – 0.2
- 0.2 – 0.5
- 0.5 – 1.0
- 1.0 – 5.0
- 5.0 – 10.0
- 10.0 – 15.0
- > 15.0

Since the general acceptance in the 1960s of the seafloor spreading concept, numerous observations have reinforced it. WHOI scientist Brian Tucholke has mapped sediment thickness in the western North Atlantic Ocean basin, showing it is thin or absent near the spreading center, where the ocean floor is young (indicated in yellow), and exceeds nine miles thickness near the continental margin, where sediment has accumulated for over 100 million years (shown in blue or red). Photo: B. Tucholke, WHOI

months in the darkness of the ocean, the sounds they give off are picked up by hydrophones at listening stations located hundreds or thousands of miles away on ocean islands and around the margin of the sea. By a kind of triangulation, the floats' positions can be precisely located, which over time reveals the complex and irregular paths of mid-water currents and eddies—again, proof that the depths of the ocean are by no means quiescent.

Seafloor Spreading

Possibly the most important development in Earth sciences in the last quarter century has been recognition of the process of seafloor spreading and confirmation of the larger theory of continental drift. Development and proof of these concepts was the result of many scientists' research, going back many decades, and gigantic joint ocean research programs.

In 1968, two marine geologists, Arthur Maxwell (now at the University of Texas) and Richard Von Herzen, a Senior Scientist at WHOI, set out with their colleagues to test the hypothesis of seafloor spreading. Using the drilling ship *Glomar Challenger*, Maxwell and Von Herzen drilled eight holes along a 1,500 kilometer transect crossing the mid-ocean ridge in the Atlantic Ocean. The holes were drilled through the blanket of marine sediment into underlying crustal basaltic rock, whose age had been estimated by another colleague, James Heirtzler. Heirtzler had indirectly dated the crustal rock using magnetic patterns in the oceanic crust, which, like bar codes used to price packages at the grocery store, contained important information—in this case crustal age.

With actual samples of the sediment immediately overlying the ocean crust, the researchers could compare dates using micro fossils of marine life with Heirtzler's independent estimates. As predicted, the result was that the ages agreed. Both crustal rock and overlying sediments got older moving away from the mid-ocean ridge. These observations indicated that the average rate of seafloor spreading had been about two centimeters (⅞ inch) per year over a period of sixty million years, and that the ocean had widened by 3,500 kilometers at this part of the Atlantic. Thus, additional evidence was provided to confirm the once ridiculed hypothesis of continental drift, even though the mechanism driving this movement (once cited as the principal flaw of the continental drift hypothesis) has yet to be elucidated quantitatively.

Scientists suspected mid-ocean ridges would be geologically active sites, from the high heat flow through the seafloor at these sites, and the concept of seafloor spreading, which identifies mid-ocean ridges as the location of oceanic crust formation. However, they were by no means prepared for the discoveries made during exploration of the mid-ocean ridges using the Deep Submergence Vehicle (DSV) *Alvin*. Scientists aboard *Alvin* came upon underwater hot springs (hydrothermal vents), spouting water as hot as 350 degrees centigrade, surrounded by communities of giant invertebrates, never before seen by man. Clouds of chemical precipitates poured from some of these vents as white or black smoke. Whole research programs have come from these discoveries, to explain how abundant life could exist in the deep sea where no photosynthetic plants exist. Results of this research showed that the communities are sustained by chemosynthesis, in which energy released by chemical reactions, rather than photo-reactions, drives the metabolism of bacteria living in the vents and the tissues of some of the larger animals. Although chemosynthesis has long been recognized as a major form of metabolism, it had never before been displayed on such a scale or in as striking an environment.

Recirculation of seawater through hot rocks at the mid-ocean ridge hydrothermal sites also opened to chemists an entirely new range of chemical reactions that could alter the composition of seawater. Once thought to result from rainwater dissolving continental rocks, the composition of seawater is now believed to be significantly affected by high temperature reactions.

Project NOBSKA

In addition to its role in ocean research, WHOI has served an important function over the years as an effective, neutral meeting ground for analysis of important ocean issues. In the summer of 1956, the Undersea Warfare Committee of the National Academy of Sciences conducted a project, dubbed NOBSKA, to examine the undersea as an arena of naval warfare. One of the salient conclusions of the project was that it would be possible to launch long range missiles from submarines, a conclusion with gigantic military, economic, and political impact for the world.

Columbus Iselin, then Director of WHOI, and Ivan Getting of Raytheon Manufacturing Company (formerly of M.I.T.) coordinated the project. About sixty permanent experts participated—naval officers, engineers, and oceanographers from the Institution itself—

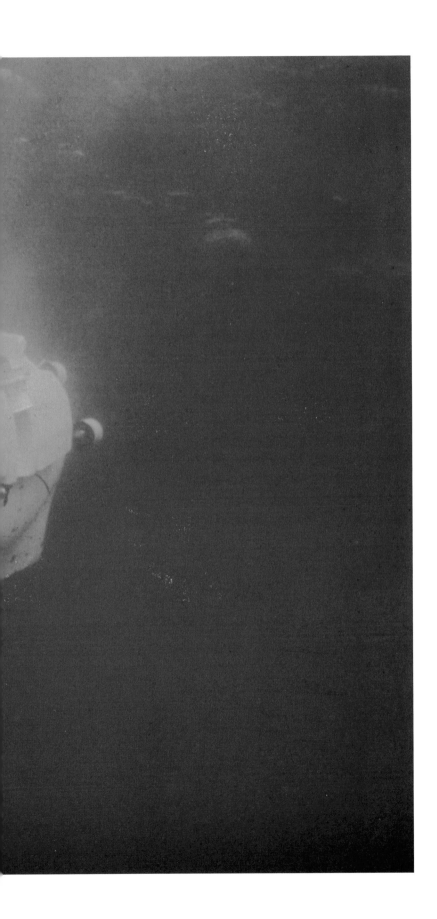

Since 1964, the DSV *Alvin*, named after Allyn Vine, has provided scientists with a unique vessel to further their explorations of the deep sea. *Alvin*'s titanium-hulled personnel sphere carries one pilot and two scientific observers to depths of 4,000 meters on 200 day-long dives each year, and remains the most active research submersible in the United States. Photo: Rodney Catanach, WHOI

Photographed from the DSV
Alvin, black mineral
precipitates spout from vents
on the seafloor along the East
Pacific Rise. The great oceanic
pressure prevents the water
from boiling, even at the
recorded temperature of 350
degrees centigrade. The
discovery of hydrothermal
vents, such as the "black
smokers" shown here,
indicates that the composition
of seawater can be influenced
by exposure to very high
temperatures. Photo: Dudley
Foster, WHOI

Special lifting equipment used
to deploy and retrieve the
DSV Alvin (shown partially
submerged at the left), is
mounted on the stern of R/V
Atlantis II. The "A-II" was
modified in 1983 to serve as
Alvin's dedicated mothership,
extending the range and
number of dives possible on a
single voyage. Photo: Jerry
Dean, WHOI

as well as a number of visiting participants. Some of these distinguished scientists and engineers, now senior level or emeritus, still live in the Woods Hole area, such as Allyn Vine, Robert Morse, and Robert Frosch.

For their meetings, the NOBSKA project leased a private estate in Woods Hole, the Whitney Estate, and family members of the participants enjoyed the summer life of Cape Cod for the three months of working sessions. This project was notable for the effective cooperation and interaction among academicians, military and government personnel, and people from industry.

One significance of NOBSKA was that it allowed expert review of an issue for which the government had already made a decision, in an open intellectual climate. Prior to NOBSKA, submarines had been viewed primarily in terms of their threat to surface shipping. As an economy measure, President Eisenhower had directed the Navy not to develop new missiles, but rather to use the Army *Jupiter*, a liquid fuel missile that weighed about 150,000 pounds and measured sixty feet in length. Clearly, deployment of such a missile by the Navy would be limited to large ships.

The startling conclusion reached at NOBSKA was assessed and confirmed by a task force supervised by Paul M. Fye (who became Director of WHOI two years later) under the Chief of Naval Operations. The first fleet ballistic missile, known as *Polaris* A-1, was deployed aboard the submarine *George Washington* in November 1960.

Given the recent reorganization of Europe and the Soviet Union we may have reason to hope superpower confrontations will become a thing of the past. However, there are many other areas in which government ocean policy needs to be independently and authoritatively examined and the range of options re-explored. This is only one area where the Woods Hole Oceanographic Institution has an important role to play in the future.

Acknowledgments

While still Director of WHOI, Dr. Paul M. Fye said, "Nobody speaks for the Woods Hole Oceanographic Institution," referring to his philosophy that at this Institution, viewpoint, creativity, and initiative arise at the level of the individual. I find the generosity and optimism and confidence of his statement moving.

My remarks about WHOI are not intended to be construed as "official" or complete. I hope they are correct. Each person who has known this place undoubtedly has a unique view of it, influenced by many factors. Ms. Laura Predario helped assemble this text. I thank Ms. Victoria Kaharl for her numerous suggestions, including that, in the end, I must decide upon the content for myself.

I dedicate this to Dr. Fye, who died in 1988. His acts of kindness and support affected the lives and careers of many of us, and his example will always be a model.

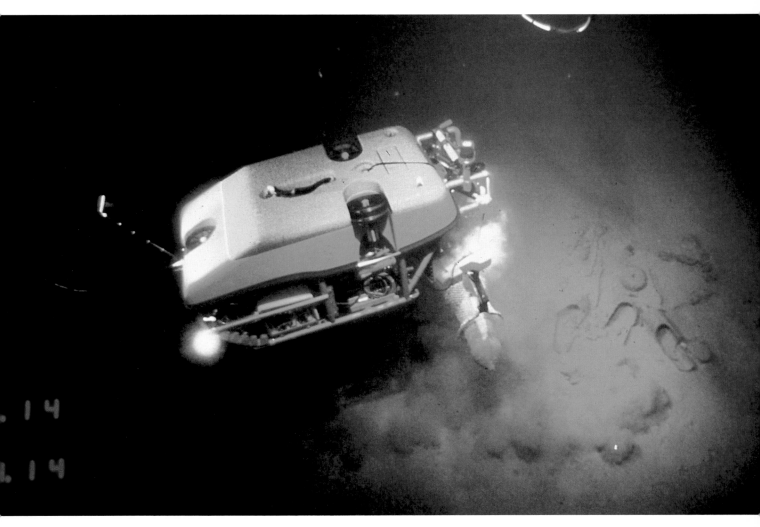

The remotely-operated vehicle, ROV *Jason*, conceived at WHOI's Deep Submergence Laboratory, illuminates ancient treasures from a shipwreck on the Mediterranean seafloor. Because of their superior dive durations, elements of safety, and lower cost, ROVs will most likely replace manned submersibles for many seafloor tasks. Video cameras on ROVs generally provide better visibility along the seafloor than is possible through the portholes of manned submersibles. Photo: Quest Group, Ltd.

3.

The Bedford Institute of Oceanography, Nova Scotia

John H. Vandermeulen and
Bosko D. Loncarevic

The Bedford Institute of
Oceanography. Photo: BIO

The Beginnings: 1497—
"A Sea Full of Fish"

"The sea there is full of fish to such a point that one takes them not only by means of a net but also with baskets to which one attaches a stone to sink them in the water!" With those words, in 1497, European explorer John Cabot astonished Europe with his discovery of the fishing ground off Canada's east coast—and with those same words he set off the race for what turned out to be the greatest fishing grounds in the whole of the Atlantic Ocean. At that time, those fishing grounds comprised a vast ocean area that could only be reached after an arduous voyage through uncharted waters, with unknown depths, and with only the vaguest knowledge of currents and coastlines. Those early voyages were highly dangerous, at times terrifying, as crews ventured into some of the world's most storm-whipped waters. These voyages were undertaken in sailing vessels and with navigation devices and charts that no sailor would accept today—they were literally voyages into the unknown.

With time, more and more knowledge of those far-off waters of the western Atlantic began to accumulate, as European captains cast caution aside and set sail for the fish-rich grounds. At the same time they obtained new insights concerning where they were, where they traveled, and as they mapped and charted their routes and their fishing grounds, the unknown vastness of the ocean slowly became the familiarity of traditional routes and charted depths. Their tools were the four basic "L's" of the science of navigation: Lookouts, Leadlines, Log, and Latitude, and their "data" were depth, speed, distance, and whatever information could be gleaned from the Mariner's Almanacs.

Today that same fishery remains one of the greatest, most exploited fisheries of the world, providing hundreds of thousands of tons of fish to markets in Canada, the United States, France, Spain, Portugal, England, and further off to the Soviet Union, among other distant ports.

That same fishery gave the impetus to develop Canada's oceanographic excellence in North-temperate waters. The early research investigations were conducted by fishery officers, and later by fishery scientists. Today, oceanographers and hydrographers, broadly specialized in many disciplines, chart and study Canada's waters and its marine resource potential in Canada's three oceans—on its east coast, the North Atlantic Ocean and Labrador Sea; along its north coast, the Arctic Ocean; and along its west coast, the North Pacific Ocean.

That aquatic domain includes the world's longest coastline and the second largest continental shelf. At the same time, John Cabot's "sea full of fish" now includes other vast marine resources, such as oil and gas and other minerals. The ocean uses now encompass transportation, ship-building, ocean industries, and tourism, among others. The modern explorers have new traditions, new tools, and are gaining new insights, while trying to unravel the old ocean's secrets.

The new ocean science is an amalgam of geology and geography, of physics, chemistry, and biology, that has led to a new fusion of sciences with new devices and methodologies—from towed electronic plankton-counting devices that automatically record their measurements for later analysis, to remote, deep-water current-meters that can return to the ocean surface from the ocean floor at a simple acoustic command from the surface ship, to sophisticated arrays of towed acoustic receivers that can record echoes from the earth's layers tens of kilometers below the ocean floor. The ecologists investigate plankton phenomena in waters from the equator to the North Pole, physical oceanographers model long-term weather phenomena in the world's first concerted effort to predict ocean storms, while geologists focus on sub-seafloor structures and processes, and biologists examine pollution of the arctic food-chain from their floating ice-islands. As for the chart-makers, the lead line is now replaced by *Dolphin* (a radio-controlled launch), and the future hydrographic charts will be complemented by computer driven video displays.

BIO—The Bedford Institute of Oceanography

Located on the shores of Bedford Basin, on the Atlantic side of the province of Nova Scotia near the cities of Halifax and Dartmouth, the Bedford Institute of Oceanography is ideally situated for a wide range of oceanographic scientific studies. Officially opened in 1962, with a major expansion in the late 1970s, the Institute is Canada's foremost federal marine labora-

The modern surveying for
harbor approaches and
inshore waters includes the
use of a survey catamaran,
Smith. The data collected can
be assimilated into
computerized electronic
packages for onboard
navigation. Photo: BIO

tory with a research fleet capable of cruising all the world's oceans. There are several smaller federal marine research centers in Canada—the new Maurice LaMontagne Institute in the Gulf of St. Lawrence, and the Institute of Ocean Sciences in Patricia Bay on Canada's west coast are two examples—but BIO remains the flagship of Canadian marine research, not only in size and scientific manpower, but also in the breadth of oceanographic science in which its scientists engage, and in the volume of data and publications that it produces annually.

BIO comprises laboratories and divisions of several Departments of the Federal Government (Fisheries & Oceans, Environment Canada, Energy/Mines & Resources, with the Department of Fisheries & Oceans acting as the Institute manager). The Institute operates five large ocean-going research vessels ranging in overall length from 50.3 to 90.4 meters, as well as several smaller vessels used primarily for inshore and coastal studies.

Besides conducting the major oceanographic programs, the Institute houses a sea-bird research unit, an environmental protection unit, a number of marine-science related private companies with a variety of marine expertise, and (until recently) the Secretariat of the Northwest Atlantic Fisheries Organization (NAFO).

The staff at BIO includes over 250 research scientists, 80 hydrographers, 200 ship crew, and 300 support staff. Each year the Institute plays host to dozens of visiting students, and scholars, as well as to Canadian and foreign scientists. In addition, many of the staff travel to other institutions, either on collaborative research programs or as participants in conferences, research meetings, or other national and international convocations and working opportunities.

BIO is not a teaching facility, and is engaged mainly in marine scientific and engineering research, resource management, and hydrographic charting. A number of the senior scientific staff also hold joint teaching positions as research associates and adjunct professors at nearby universities, especially Dalhousie University in Halifax. This practice is encouraged by the local marine scientific community, which argues that such government-academia interaction acts as a strong intellectual stimulus for joint research and graduate study programs. Many of the scientific staff also participate in various international oceanographic teaching and training programs and, from time to time, are seconded to international organizations and interdisciplinary programs. Also, laboratory facilities and ship time are provided for graduate

students from several Canadian east coast universities. Visiting scholars and graduate students have come from as far as the West Indies, South America, Southeast Asia, and China to carry out marine studies in Canada.

The A.G. Huntsman Silver Medal for Excellence in Marine Sciences was established at the BIO in 1980 in recognition of the fundamental and seminal work carried out by oceanographers around the world. Recipients have included marine scientists and oceanographers from Spain, the United States, France, and the United Kingdom.

Multi-Disciplinary Oceanographic Research

The intimate coexistence of different research disciplines, together with a large central library and a fully integrated computing facility, has given the Bedford Institute of Oceanography its deserved reputation as a world center for interdisciplinary marine science. This is reflected in the international nature of programs and ocean cruises that have involved BIO staff, and in major interactive programs as, for example, the Grand Banks and George's Bank programs; WOCE (World Ocean Climate Exercise); Canada-Peru ocean studies cruises; joint programs with other institutes such as collaborative studies of the Mid-Atlantic Ridge with the Woods Hole Oceanographic Institution; studies in deep ocean ecology; ocean pollution research programs; participation in the International Ocean Drilling Program (ODP), and such ventures; the arctic ice-island research station; the development of *Dolphin*; and BIO's special area of expertise in sensor technology, data acquisition, and data processing.

In the present account we cannot cover the full range of activities at BIO. Instead, we have selected a number of major projects to illustrate the work of our Institute. These were selected either because they included large multidisciplinary teams or because their scope required a commitment over a period of years.

Integrated Oil Pollution Studies

Oil pollution became a public concern only in the late 1960s with a seeming wave of tanker incidents beginning with the sinking of the tanker *Torrey Canyon* in the English Channel. This was followed shortly by the *Florida* spill (1969), the *Arrow* spill (1970), the *Argo Merchant* spill (1976), off the coast of France, and that of the

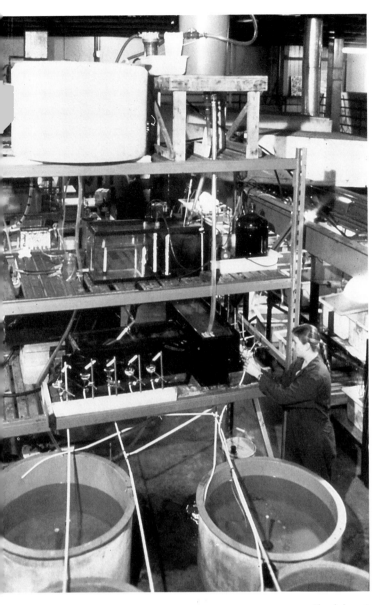

The laboratories of the
Bedford Institute of
Oceanography are equipped
with up-to-date scientific
instrumentation to carry out
research in all marine
disciplines. Photo: Kelly
Bentham

The A.G. Huntsman Award was established by BIO to recognize excellence of research and outstanding contributions to the marine sciences. The award, presented annually since 1980, honors those men and women, of any nationality, who have had and continue to have a significant influence on the course of marine scientific thought.
Photo: BIO

Exxon Valdez grounding and spill off the coast of Alaska in 1988.

Questions about the fate and effects of oil spills probably have generated more interdisciplinary investigations than any other problem in oceanography. Before the *Torrey Canyon* spill, concern over oil in the oceans was voiced primarily by the tourist industry. Tar balls, from the cleaning and flushing of tanker holds at sea, were finding their way onto beaches around the Mediterranean, Bermuda, the Bahamas, and other places whose economies were dependent on steady income from visitors. With the *Torrey Canyon* and subsequent spills, a new scientific effort to understand "the fate of the oil" in the marine environment began in many laboratories. Only gradually was it realized that new questions must be asked about the ecological impact and physiological properties of oil and its various hydrocarbon constituents.

Beginning with the spill and release of oil into the seas and onto the sea surface and moving to questions about ecosystem impact, spill research combines the methods and science of hydrocarbon chemistry, physical oceanography, physical chemistry, coastal and sediment geology, fisheries and intertidal biology, and microbiology. All these are necessary, and must be integrated to understand the spreading of oil slicks, mixing of oil and the water column, evaporation of the volatile components and the resulting "weathering" of the spilled oil, as well as the inevitable process of oiling of shorelines and sub- and inter-tidal animal and plant communities. Added to this is the new science of environmental toxicology or ecotoxicology that focuses on the sensitivity of marine organisms to new, unknown toxic and possibly mutagenic components, never before examined under this new light.

On February 4, 1970, the 18,000 tons (deadweight) Liberian registered tanker *Arrow*, with a cargo of 108,000 barrels of heavy Bunker-C oil, ran aground on charted Cerberus Rock in Chedabucto Bay, Nova Scotia. By February 12, approximately 2.5 million gallons of the viscous black oil had escaped, to pollute the waters of the Bay and the surrounding shoreline. Part of the oil escaped to the open sea, some landing on the beaches of Sable Island, 100 miles from the source. Of the 375 statute miles of shoreline in the Bay area, 190 miles eventually were contaminated to various degrees. Canada had experienced its first major oil spill.

A high-level Task Force was formed quickly to coordinate the response of various agencies to the new challenge. The Director of BIO, Dr. W.L. Ford, was appointed as the scientific coordinator responsible to the Task Force. He mobilized the scientific expertise available among the Universities and Laboratories in the Halifax-Dartmouth area and during the following months up to sixty BIO staff members participated in the work of the Task Force.

The problems tackled by this scientific team ranged from the immediate to the long-term: How to remove almost a third of the cargo trapped in the sunken portion of the tanker; how to clean the fishermen's nets and boats; toxicity and taste of lobsters which may have ingested oil; study of various chemical and mechanical means of cleaning the beaches; study of the damage to the marshes and other habitats of wild birds; detailed descriptions of the currents and wind driven circulation in the area in order to predict and follow the dispersion of the spill; the wave dynamics, erosion, and re-deposition along the shorelines; and the long-term chemical and structural changes of the oil residue. The immediate results were the invention of the "slick-licker"—a device for skimming the oil from the surface of the sea now used worldwide, and the design and construction of steam cleaning equipment for half a mile long herring seine nets. The longer term investigations led to a new understanding of the ocean circulation on the Nova Scotia shelf (of importance for the regulation of the oil exploration industry); the discovery and tracking of tarballs along main ocean routes and their dispersion by ocean currents; new modeling of beach erosion mechanisms; and many other discoveries, which help us evaluate the damage to the environment caused by marine oil spills.

Another result has been a new, interdisciplinary way of looking at, and understanding, the resilience of marine communities. Work at the major oil spill sites suggests that marine ecosystems are remarkably resilient, to all intents and purposes recovering within a few years from even the grossest oiling conditions. The answer to this seems to lie in part with the remarkable self-cleaning powers of the oceans themselves, eroding and decomposing stranded oil and tar from the shorelines through wave action and sediment movement. In part, the answer is also in the life cycle of most intertidal organisms—a planktonic or larval stage followed by the sedentary adult stage. Thus, over time, even the most heavily impacted shoreline will become revegetated with kelp and recolonized with shore fauna.

Continuing oil pollution studies at BIO include long-term monitoring of the 1970 *Arrow* spill site, physiological studies into oil caused stress in fish, and the role of residual petroleum hydrocarbons in benthic metabolism. Field research focuses on long-term

Video recordings show the
oiling of coastlines after the
spill from the grounding of
the tanker *Arrow*. Photo: BIO

weathering and the persistence of remaining tar and oily beach sediments (of special importance in this cool, north-temperate environment), on the mechanisms of oil entrapment in coastal sediments, and on the long-term biological activity of these oil residues upon intertidal organisms. Recent findings indicate that these may still retain some capability to induce mutations in the laboratory bacterium Salmonella typhimurium, suggesting that oil spills may actually have a longer lasting subtle ecological impact than the gross mortalities and oiling seen immediately after the spills. However, other studies suggest that some petroleum hydrocarbons may in fact become metabolized by certain micro-organisms and may well serve as a source for their energy.

Geoscience Mapping

The end of hostilities after World War II created new opportunities for exploration of the offshore areas. Surplus navy ships were available for civilian use (two such ships were used by BIO till the late 1960s). Improvements in electronic positioning (*Decca* and *Loran*-C and more recently navigational satellites) and other equipment developments made accurate and efficient offshore surveys possible. Finally, new national interests in better fisheries management, in the evaluation of mineral resources, in greater safety for vessels that were rapidly increasing in size and tonnage, and in defensive measures to combat a perceived submarine threat, all required better maps and up-to-date information concerning the morphology and structure of the seafloor.

The Canadian Hydrographic Service (CHS) is responsible for the production of navigational charts along carefully controlled, closely spaced lines. Immediately upon the founding of BIO, these surveys were recognized by the geologists and geophysicists as an ideal opportunity to establish a program of geoscience mapping. As carried out today by the Atlantic Geoscience Center of the Geological Survey of Canada, this program comprises a wide range of activities, five of which will be briefly discussed here.

Joint multi-disciplinary surveys with CHS (supplemented more recently by aeromagnetic surveys) have been a continuing program since 1964. This program is perhaps the most comprehensive survey in the world considering the vast area to be covered. The project has produced many different types of maps as a contribution to the description of the Canadian landmass. The methodology is exportable and has been

used, for example, on a survey of the continental shelf of Senegal under the sponsorship of the Canadian International Development Agency (CIDA).

The data collected during these surveys have enabled BIO geophysicists to construct a detailed continental drift reconstruction of the plate motions between the North American and Euro-Asian Plates, showing a complicated series of stretching and break-ups along different portions of the continental margin. The data have also been used to demonstrate the extension of major geological trends, such as the Appalachian mountains, from land to offshore areas. On a larger scale, the data are used for the evaluation of the International Geomagnetic Reference Field, and for comparisons between the satellite derived gravity field and sea-surface measurements.

Surficial and bedrock mapping, as conducted by BIO geologists, uses echo-sounding records, high resolution seismic profiling, side scan sonar and sampling techniques. In recent years, multi-beam and deep towed seismic equipment have been added to the list of the geologist's tools. The output of this program has been a series of maps showing the geological characteristics of the sea bottom, as well as the age and type of underlying bedrock. These maps are used by fishermen to improve the catch efficiency, by telephone companies to select the routes for underwater cables, by the mineral industry to assess the potential sites for the dredging of aggregates, and so on. This mapping also revealed the offshore extent of the last ice sheet and its movements. One interesting discovery was the presence of small vents on the seafloor, called "pock marks," caused by escape of methane and other gases formed by the decomposition of organic matter. More recent work on the continental slope has detected large-scale slumping of unstable slopes and mapped a large slump responsible for the devastating earthquake of 1929.

Deep sea exploratory mapping: With a huge, unexplored continental shelf under Canadian jurisdiction, the deep sea mapping could be only a limited effort. In addition to work done under the auspices of the *Hudson*-70 expedition (see below), two examples of deep sea exploration will be mentioned: Mid-Atlantic Ridge (MAR45) at 45°N and Labrador Sea ODP Site surveys. The MAR45 was a pioneering effort when initiated in 1965 and it set an example for the French-American project FAMOUS and subsequent ODP Site surveys. Formulation of the Vine-Matthews hypothesis of seafloor spreading in 1963 focused the attention of marine geophysicists on the chain of mountains extending down the center of the North and South

Atlantic. BIO was the first institution to mount a concentrated exploration of a small portion of this ridge, a one-degree-wide strip between 45° and 46°N. The results of four expeditions showed that the youngest rocks were in the central valley and increased in age toward the flanks; that the predominantly basaltic rocks extruded from hot magna were subsequently replaced by serpentinite, which caused the appearance of magnetic anomalies; that the speed of the seafloor spreading was not constant; and that the crustal rocks were thin under the central valley, and rapidly increasing in thickness away from the crest.

As a member of the International Ocean Drilling Project (ODP), Canada was a sponsor of several exploratory holes in the deep Labrador Sea and Baffin Bay. Prior to the drilling, site surveys were required to ensure the safety of drilling and to delineate the scientific questions to be answered by drilling. Using techniques developed on previous surveys, CCS Hudson in 1984 carried out surveys of three sites, two of which were subsequently drilled. The samples recovered showed that the paleo-circulation of ocean currents was influenced by the sequence of continental break-up between North America and Greenland. The dating of micro-fossils directly above the basement basalts gave an estimated age of fifty-five million years, thus confirming the history of ocean floor spreading in the Labrador Sea, as previously deduced from the analysis of magnetic anomalies.

The Basin Analysis project uses reflection seismic data, mostly supplied by industry on a confidential basis under the Canadian regulations for offshore exploration. This project produces detailed maps of the structure, age, and types of sediments in the basins on the continental shelf and slope. On the basis of these maps the project produces a revised estimate of the national reserves of oil and gas. This important information is used by the Government to formulate national energy policies, and by industry in determining their exploration strategies.

Deep seismic profiling is an extension of the above Basin Analysis using geophysical contractors to acquire seismic profiles that penetrate many tens of kilometers into the earth's crust. The data are used to add the third dimension to regional geological mapping. The results of this five-year-old program have shown that the crustal thickness is different under different geological provinces and depends on the collision of continents that produced the roots of the Appalachian mountains. Of particular interest is the existence of deep faults which, in places, cut through the whole thickness of the crust to the underlying

CSS Hudson is the flagship of the Canadian oceanographic fleet. She has a full ice-breaker hull and is capable of working in all ocean climates and sea states. Photo: BIO

Biological survey cruises of the Scotian Shelf, George's Bank, and the Grand Banks provide the necessary data for marine ecological modeling and fish stock evaluation. Photo: BIO

mantle. These faults are the boundaries of the offshore sedimentary basins formed by the stretching and extension of rocks in the early stages of continental break-up, which led to the formation of the Atlantic Ocean and Labrador Sea.

Coastal Ecosystem Biology

The Grand Banks of Newfoundland represent an immense fishery resource with complex fishery stock implications. This vast resource is being exploited by a truly international fleet. However, as it falls largely within Canada's 200-mile economic zone, a considerable part of the BIO scientific effort is directed towards unraveling the number of environmental and human processes that influence this unusual coastal ecosystem.

Biological, physical, and chemical oceanographic sampling cruises to the area over the last decades have provided a broad data set on larval fish recruitment, age-groups of different fish species, plus myriad observations such as nutrients, seasonal and depth profile temperature readings, salinity measurements, SPM (suspended particulate material), turbidity data, and observations on light penetration and attenuation.

It appears that the Grand Banks have some unique features, not the least of which is that the classical diatom-copepod system (the former being the primary, and the latter, the secondary production) holds for only the spring of the year. For the remainder of the year the system consists of a micro-phytoplankton/micro-zooplankton food chain, supported by a secondary food-net of bacteria depending on the micro-phytoplankton production.

Paralleling these biochemical and micro-ecological studies, other fish stock assessment studies address the upper part of the Grand Banks food chain. Done in collaboration with fishery scientists from BIO's sister organization in Newfoundland, these emphasize understanding the process of recruitment of larval fish into the principal stocks, year-class strength, and the movement of fish stocks over the Grand Banks.

Measurements by physical-chemical oceanographers show the physical transport of nutrients from the deep waters up onto the shallow sunlit waters of the Banks where the supplies of nutrients play an important role in the survival of fish eggs and their development into free-swimming demersal juveniles.

The interdisciplinary study of the Grand Banks was inspired by the hypothesis of former BIO scientist

Dr. Bill Sutcliffe and colleagues, which says that the food chain of the Grand Banks derives from the transport of nutrients from high surface nutrients of the more northern Hudson Strait. This hypothesis continues to be tested as more data come in from the Banks, and from the waters of the Labrador Shelf and the Hudson Strait—a study in ecosystem productivity spanning thousands of kilometers.

Modern technology, in the form of mathematical modeling, has brought new insights into the interpretation of the columns of new data. The Grand Banks modeling project represented the first attempt at devising a whole ecosystem model that included all the various biological and environmental components—the latter as more detailed submodels to the overall model. As is the case with various modeling attempts, in areas that are poorly understood, many gaps in the understanding of the system were identified. For example, the interplay between nutrient dynamics and primary production, particularly the role played by microbes, was not well understood. Another area where such information was lacking was in the linking between the pelagic components of the ecosystem and the benthic organisms. Feeding rates, in the ocean environment, were another gap.

While the modeling effort perhaps represented more of an exercise than a solution in understanding the Grand Banks fishery, the multi-year and multi-disciplinary investigations are a significant step in synthesizing the numerous data that constitute the output from a vast and complex oceanic ecosystem. Similar studies have now involved an ecosystem description of the nearby Bay of Fundy, famous for its great tidal ranges and its high productivity (mariculture, pelagic fishery, shellfish fishery). Also, first attempts are being made at modeling the vast resource system of George's Bank, one of the world's richest continental shelf areas, which has long been of interest for its suspected equally rich oil and gas reservoirs.

Hudson-70

The circumnavigation of the Americas by BIO's research vessel C.S.S. Hudson, in 1970, ranks as one of BIO's grander efforts of discovery, and of logistics. The expedition, in the style of other old and famous expeditions—the Challenger, Discovery, Meteor, and Vema—combined the elements and skills of many oceanographic sub-disciplines to sample the waters, currents, seafloor, and ecosystems of the oceans surrounding North and South America. The data collected

on that eleven-month long oceanographic experiment, the first circumnavigation of the two continents on a single voyage, enabled oceanographers to compare, almost simultaneously, oceanic ecosystems and processes separated by thousands of miles and under vastly differing climatic conditions.

The results and new findings from this expedition, reported in dozens of scientific publications, still influence current thinking about global oceanic processes. One of these findings, by Dr. Ray Sheldon, was that in the oceans material occurs at all particle sizes in roughly equal concentrations, within a range from one micron to about 10^6 microns, i.e., from bacteria to whales. A second finding was the discovery that the earth's crust underlying Baffin Bay is of oceanic and not terrestrial character, thereby defining the opening and spreading history of continental drift between Canada and Greenland. A third finding was that of "pingoes" on the floor of the Beaufort Sea, geological features more normally associated with the arctic landmass. A fourth finding was the discovery of the scouring of the seafloor by winter ice, which was immediately recognized as a serious potential hazard to pipelines and any other seafloor installations associated with polar oil and gas production.

One finding, that the oxygen concentrations in the waters of the South Atlantic had remained essentially constant over the forty years preceding the cruise, was considered primarily of general interest at the time. Today, this finding has important implications for predicting global change.

On this voyage "round the Americas," the Hudson transited the famed ice-packed North-West passage through the Beaufort Sea, the world's first oceanographic research vessel to achieve that distinction. The geological and geophysical investigations in the Beaufort Sea and Baffin Bay contributed to the awakening interest in the hydrocarbon potential of the Canadian arctic offshore.

Dolphin and Larsen

Brainchild of BIO marine engineers, Dolphin (Deep Ocean Logging Platform with Hydrographic Instrumentation and Navigation) is to provide answers to ocean-mapping questions that began with John Cabot—that is, how to map the vast areas of ocean floor most efficiently and effectively, especially in the open, rough sea conditions of the North Atlantic. The solution is a fleet of unmanned, snorkeling mini-submarines that travel just below the sea surface,

Estuarine research combines
the talents of oceanographers,
biologists, benthic ecologists,
and mathematical modelers.
Field studies on intertidal
mud flats provide data on
benthic production in the Bay
of Fundy. Together with
studies of fish productivity,
physical oceanography, and
sea-bird ecology, these allow
examination of the complex
interactions of large coastal
ecosystems. Photo: BIO

Several hundred icebergs are carried by the cold Labrador current southward along the Labrador and Newfoundland continental shelf. When stranded on the shallow banks of the shelf they cause seafloor scouring, which can be viewed with side-scan sonar. Photo: BIO

thereby avoiding the tremendous energy dissipation in the wave zone. Each is equipped with a depth-sounding system and a radio link for remote control from a master station on the mother ship. An array of such *Dolphins* are envisioned as spreading out on either side of a mother vessel, with the capability of sweeping vast stretches of the continental shelf, while radioing their information back to the central computer.

In a broader context, *Dolphin* will provide additional means to meet the demands of coastal states for ocean-mapping as a result of the new Law of the Sea requiring better descriptions of continental shelves and offshore areas.

Larsen takes the hydrographers into the space age of air-borne laser-mapping using a scanning laser to measure water depths from a low flying aircraft. The technique is being investigated for its potential usefulness under Canadian sea-state conditions. Both *Dolphin* and *Larsen* will feed their information into computerized data management systems, that will ultimately lead to electronic charts to satisfy the needs of everyone who does business at sea, be they fishermen, oil companies, the transporters of cargoes, or warships.

CASP—Canadian Atlantic Storms Program

The ocean responds readily to atmospheric conditions at its surface. For example, wind plowing over the sea generates waves on its surface, from ripples less than a centimeter high, to storm-driven waves that can reach more than 15 meters in height. Wind can also reach further down into the ocean, causing motion well below the sea surface affecting the physical structure of the water column, but also influencing the distribution of fish egg masses and the phyto- and zooplankton. Fluctuations in sea level can also be caused by surface wind—in some parts of the world causing wreckage, devastation, and even deaths because of storm-related "storm surges." Surprisingly, a model of nontropical storms first developed by Norwegian meteorologists in the 1920s continues to be used for storm prediction today—the familiar low and high pressure systems of television and newspaper weather maps finding their origins in that early mathematical depiction. That model, however, describes only the so-called large-scale features of storm systems—thereby missing the myriad of fine detail that constitutes the fronts and precipitation zones.

Recent research, showing that small-scale features also exist within storms, led meteorologists from

A hydrographic survey
platform, *Dolphin* can be
remotely controlled from the
mother ship within a range of
about twenty kilometers. A
fleet of these platforms can
multiply the effectiveness of
offshore surveys many times.
Photo: BIO

Canada's Atmospheric Environment Service and BIO oceanographers into a collaborative experiment to obtain detailed descriptions of storms in a way never seen before. The Canadian Atlantic Storms Program (CASP) succeeded in this by using a network of observers, land- and sea-based automated instrumentation, and recording instrumentation launched by balloons in the four east-coast Atlantic provinces to record surface pressures, temperatures, wind and precipitation, humidity, cloud patterns, as well as ocean currents, waves and water properties. The latter were made every half hour over a four-month period, generating thousands of data sets for later computerized analysis.

The combined study enabled an especially detailed description of the storm winds that affect the Scotian Shelf, yielding one of the largest and most complete sets of wave data ever gathered. It also provided insight into the rate of growth of waves due to the wind in the early stages of their development, and contributed to a more complete understanding of winter storms on Canada's east coast. The insights and theoretical principles gained through CASP will ultimately be of value in the study of storms in other regions of the world's oceans.

A tangible outcome of the program was the discovery of the presence of continental shelf waves—large, low-frequency waves that are generated in continental shelf regions in response to severe storms. With periods of days and length scales of hundreds of kilometers, they move along the coast and presumably affect the entire shelf ecosystem by displacing nutrients, egg masses, and plankton.

The Arctic

The Bedford Institute of Oceanography has launched oceanographic investigations into Canada's Eastern Arctic since the 1960s, at first focusing on geophysical problems but today including all aspects of arctic marine science. In the early days, the investigations were often contests of hardy men against the environment. Today, upon occasion, those investigations may still call for roughing it in very spartan field camps, often staked directly on the ice, but they also include modern communication and can employ the most modern means of transportation—from research ships and helicopters to semi-permanent laboratories located on floating ice islands.

Fixed base camps are also part of this effort, offering accommodation, landing strips for fixed-wing aircraft, garages, and staging areas for field expeditions. The Canadian-designed and -manufactured Otter aircraft, and its later equally famous cousin, the Twin Otter, have been the work horses in this remote region. Equipped with turbine engines, a sophisticated wing design, and fitted with low-pressure tires, they can land and take off at unprepared landing sites, and haul in 1,000 kilograms or more of food, supplies, and scientific equipment. High-altitude radio and special satellite-linked navigational systems provide precise locations of observation points. However, work in the High Arctic involves a very high cost, and research programs inevitably include several research institutes and agencies, combining their multidisciplinary talents and logistical skills.

Arctic Biological Oceanography

Arctic studies by biological oceanographers at BIO have focused on the marine food chain—the free swimming phytoplankton and zooplankton, and the "epontic" community in and just under the ice—in an attempt to find the driving forces of what must be one of the most intense growing seasons in the world's oceans. Because of the remoteness of the field stations from laboratories located in Dartmouth, N.S. (air distance of over 3,000 kilometers, requiring up to ten days transit time by ships), special preparations must be made to ensure that logistic problems of supply and equipment will not hinder the around-the-clock shipboard measurements.

Much use is made of "multi-leg" cruises in which the research ships *Hudson* and *Baffin* will be away from home port for three to four months at a time. Various portions of the cruises are shared by hydrographers, geochemists, chemical oceanographers, geophysicists, and biologists for periods of a few weeks at a time. As many as thirty scientists, involving as many as ten different research programs, can be accommodated on these multi-purpose cruises, flying into Resolute Bay or as far north as Thule (Greenland) to meet the ship for their scheduled leg of the cruise. Even containerization has now entered into scheduling the work of these scientists. Back at the BIO, steel containers the size of small railway box cars are fitted out as portable laboratories, containing all electrical and water services and equipped with all the necessary field instrumentation. These dedicated laboratories are secured to the decks of the research ships, where they are spot-welded into place and transported to their sampling destinations in the High Arctic. At the end of the field season, after many months at sea, these container-labs return to the Dartmouth dockside for offloading.

The focus of arctic research is on unraveling the biological production cycles of the arctic waters, dependent on such factors as day length and temperature extremes, and on questions such as regulation of growth, the unique adaptations of arctic marine organisms, and their food sources. Special shipboard photosynthesis experiments, utilizing light sources mimicking the wavelengths of light penetrating the arctic waters, are aimed at understanding the photosynthetic efficiency during the arctic's short growing season. Deep-water sediment traps measure the drift and content of oceanic particles.

More recently, the interest in arctic research has expanded to include the "epontic" or under-ice communities of algae and small invertebrate organisms. Embedded in and clinging onto the underside of ice several meters thick, production in this unique community of minute organisms comes to a peak weeks to months before the break-up of the arctic ice mass and the onset of the plankton growth in the water column. It is this incongruity in energy production that suggests some kind of link between these two production cycles—the epontic community functioning and peaking during the long ice-covered months, the planktonic community producing its high-intensity burst during the short summer season—in which the two systems complement one another.

An important conclusion from this work is that it is not necessary to invoke unique metabolic mechanisms to understand the arctic data. Rather, the arctic ecosystem should be regarded as lying at one end of a spectrum of environmental change, within which the metabolic responses change in a progressive manner.

Arctic Pollution Studies—PCBs

While generally considered pristine and unaffected by man, the arctic also is subject to our pollution and is not immune to the results of our activities that contaminate the world around us. So PCBs (polychlorinated biphenyl compounds) have become part of the contaminant loading of our polar icecap—probably transported there largely through atmospheric routes. There are, in fact, various routes for these poisonous, man-made chemicals to enter this ocean—from atmospheric sources (rain, snow, and particle fallout) river input, or through water exchange with the North Atlantic and Pacific Oceans. But the atmosphere is probably the major supply route for these semi-volatile chemicals. They originate in more southern regions, vaporizing following application as pesticides in agricultural and urban areas, or released by municipal and industrial incinerators. Whatever their sources, their measurements in the arctic environment will help us to understand their long-distance movements and transport, and will help to explain how Canada's native people in this frozen north can accumulate these chemicals through their diet when no local sources exist.

Canada's Ice Island provided a unique opportunity to locate a sampling station, while slowly drifting through the Arctic Ocean. Ice islands like this have been used by the United States and by Russian scientists since the 1950s for other scientific expeditions. Essentially immense slabs of ice calved off the edge of the Ward Hunt Ice Shelf at the north end of Ellesmere Island, they generally drift in a clockwise direction under the influence of arctic surface currents.

Measuring pollutants in the arctic environment involves a combination of hardy inventiveness and a critical eye on the chemical methodology. The thick ice mass of the ice island necessitates melting an ice-hole to reach the underlying water column, while the anticipated very low contaminant levels requires the establishment of a separate remote sampling camp about two kilometers distant from the main base camp, in order to avoid contamination from that direction.

Sampling of air, snow, and ice, proceeds together with collection through the ice hole of phytoplankton and zooplankton samples and of water and bottom sediments. Not all analyses are complete, but several specific chlorinated compounds have been identified. These include hexachlorocyclohexane, hexachlorobenzene, chlordane, DDT and dieldrin—all used as pesticides in Asia and North America. Concentrations in the arctic water columns are generally very low (1 to 10 picograms per liter), but these are high enough to be measured accurately. Interestingly, the concentrations noted in melt water obtained from snow and ice are similar to levels found in the near surface water (500 to 5,000 picograms per liter), suggesting that the source of these compounds is probably atmospheric rather than by influx via ocean currents. As with other contaminants, benthic organisms (amphipoda) apparently concentrate the less water-soluble pesticides and PCBs in their tissue lipids by factors of 100 to 1,000 times over levels present in the water.

The Arctic Ocean therefore may seem remote from industrial and human input, but global atmospheric transport makes the world a much smaller place than it once seemed.

Research on the surface of the frozen Arctic Ocean often means establishing special field camps. The rigor of working in the arctic environment is in sharp contrast to the relative comfort of modern shipboard studies. Photo: BIO

Scientists haul out a large ice
block. Metabolic studies of
the unique micro-fauna and
flora found on the undersides
of such ice blocks are aimed
at understanding biological
productivity during the arctic's
short summer growing season.
Photo: BIO

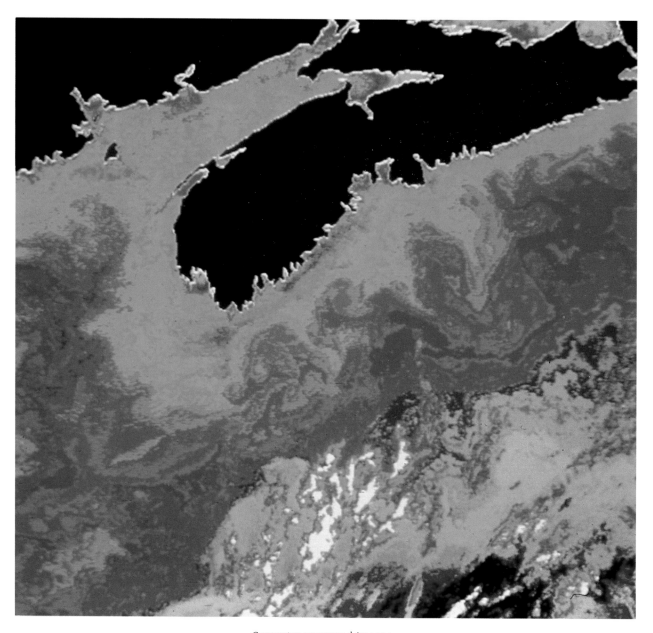

Computer-processed images
of data collected from
environmental satellites are
used to produce maps of the
distribution of sea surface
temperatures and other
oceanographic parameters.
The strong interaction
between cold northern waters
and warmer Gulf Stream
waters on the Nova Scotian
shelf is the main reason for
the rich fisheries in this
region. Photo: BIO

Future Outlook

On the practical side, oceanography will continue to captivate man's imagination with its exploration of the ocean depths and their resources. New technologies, based on sophisticated remote sensing, telemetry, and data recording will require new computerized data analyses and possibly even computerized interpretation. Already in place are the continuously recording oceanographic instrumentation packages that remain in passive remote isolation on the ocean floor or float suspended in the water column—silently recording and transmitting the data on ship- and shore-based receivers. Strings of millions of data sets will then require vast analytical procedures. These studies will undoubtedly require complex international and inter-institute programs and cruises.

Fish stock assessment is obviously at the top of the list of decided requirements, but also pressing for answers are questions about global warming, as affected by ocean temperatures. The quality of the environment is one component of the new awareness expressed under the general title of "Global Change." For the first time in the history of the Earth, man can alter the environment with, as yet, unpredictable consequences. The most drastic aspect of this Global Change may be the effect of our industrial activities on the climate with a likely warming period due to the "greenhouse effect." The oceans act as a "fly-wheel," moderating the extreme and rapid fluctuations of the weather. The relationship between oceanic processes and the resulting weather is poorly known. It is an area where new understanding must be rapidly developed, since the time scale for action on the allowable quantities of carbon dioxide released into the atmosphere may be much shorter than is generally thought—on the order of years rather than centuries.

The increasing world population and the more abundant emissions of waste products from rapidly industrializing societies threaten the quality of the environment. The oceans are the ultimate depository of many waste products, acting as a buffer and removing the toxic and other undesirable products of our civilization. But the ocean's capacity is not unlimited, and there are indications that in some situations we may be approaching the absorption limits. The result will be a sharp diminishing of our oceans' ability to sustain life. The major challenge for twenty-first century marine scientists will then be to understand the ocean's absorptive capacity and ultimately to control the type and quantity of waste byproducts that are allowed to reach the oceans. The effort presently expended on marine studies in this area is probably inadequate. As the public awareness of the dangers to which our environment may be exposed develops, the pressure of public opinion will require an expanded effort by a new generation of investigators.

Finally, as the world population approaches ten thousand million early in the next century, the ever greater need for resources to sustain that population will increase exploitation of food and mineral resources of the seas. The indications are that the seas are already less "full of fish" than they were in Cabot's days, five hundred years ago. The great challenge will be to develop management procedures that will allow sustainable catch at the optimum level, while a substantial part of the energy fuels for the new century will be drawn from offshore hydrocarbon resources.

In all these activities, future BIO personnel will have an important role to play, contributing to the well-being not only of future Canadians, but of the whole world community of nations. These future activities will be facilitated by an accelerated acceptance of the new Law of the Sea framework.

4.

The Oceanographic Museum of Monaco

Eric L. Mills and
Jacqueline Carpine-Lancre

Monaco-Ville in 1988, still
much as Prince Albert knew it
in 1910 when the *Musée
Océanographique* was
inaugurated. Photo: Anne H.
Mills

Nearly a million visitors a year crowd Avenue Saint-Martin in the Principality of Monaco to visit the *Musée Océanographique*, which, since the first decade of this century has dominated the seaward-facing cliffs of Monaco-Ville. Rising eighty-seven meters above the sea, the Museum represents the aspirations of an unusual man, Albert Ier (1848–1922), Sovereign Prince of Monaco, who made important contributions to oceanography in his own right, but who has the less usual distinction of having become a patron saint of oceanography, joining Charles Wyville Thomson, John Murray, and H.M.S. *Challenger* in the shrine of oceanographers' devotions. Despite this, Albert Ier and the origins of his Museum are not well known, and in modern oceanographic science the Museum is better known for its aquarium, as well as for its recent director, Jacques-Yves Cousteau, and his silent world below the waves, than for its contributions to science. This chapter represents an effort to place the *Musée Océanographique de Monaco* in a historical context, and to describe its activities.

Speaking in 1904 at the Sorbonne (while his museum was being built on the Mediterranean coast), Prince Albert expounded his belief that "a study of the sea will expand our intellectual domain, provide new resources for the work and well-being of human society, and give civilization a moral direction conforming to the teaching of science, reason, and justice." Six years later, addressing a distinguished audience celebrating the inauguration of the Museum, he restated his belief in the quasi-religious power of science.

> Here, gentlemen, you see that Monegasque soil has given rise to a proud, inviolable temple, devoted to the new divinity that reigns over knowledge. I have devoted all my strength of mind, of my conscience and of my sovereignty to the enlargement of scientific truth, the only ground on which the elements of a stable civilization, safeguarded against the foibles of human law, can mature.

Near the end of his life, in 1921, when he received the Agassiz Medal of the United States National Academy of Sciences for his contributions to oceanography, the Prince returned in a darker mood to the obligations and necessities of science in a world between wars. In addition to a review of his work, he desired:

> . . . to bring to you the message of our men of science, who are trying to lead the world towards a greater measure of civilization, struggling against the inherited influence of past generations of ignorance. It is a cry of revolt against those peoples of Europe whose nature, filled with ill-repressed savagery, would spread over our efforts a stifling atmosphere of imperialism and militarism.

These words, striking unfamiliar sparks in a more pessimistic age, mark Prince Albert as an unusual figure in the history of the marine sciences. His Museum shares with the Prince an anomalous but significant place in their development.

Albert Ier of Monaco, Albert Honoré Charles Grimaldi, was born in Paris, where his early education took place. Like many of his relatives, he went to sea, first in the Spanish navy as an ensign in 1866. Promoted to lieutenant, he left Spain two years later upon the overthrow of the Spanish royal family, and joined the French navy in time to see action in the Franco-Prussian war of 1870.

An enthusiastic sailor, Prince Albert bought a 200 ton yacht, *Pleiad* in 1873, renaming it *Hirondelle*. After a decade of exploratory cruises, his interest had begun to turn toward science by the mid 1880s as the result of his acquaintance with Alphonse Milne-Edwards (1835–1900) of the *Musée d'Histoire Naturelle* in Paris. In 1884, Albert was in the Baltic where he made a series of plankton collections. It appears to have been Georges Pouchet (1833–1894) who suggested to the Prince that *Hirondelle* could be used for systematic studies of ocean currents.

In 1885 Albert began a series of cruises to the central and equatorial Atlantic, releasing floats to determine the direction and speed of the currents. His scientific fame was assured by the time his first circulation charts of the North Atlantic were published in 1892.

During the thirty years after 1885, Prince Albert and an increasing naval and scientific staff spent nearly every summer at sea, progressing from current studies to midwater and deep-sea benthic trawling, fish and invertebrate traps, the capture of whales and dolphins, and meteorological work. New demands required better ships: in 1891 Albert took delivery of a new ship, Princesse-Alice, a 650-ton auxiliary steam yacht, equipped with the latest oceanographic gear including steam winches and a sounding machine, as well as a freezer, special laboratories, and electric lighting. Working from Madeira to the Azores, along the European coasts and in the Mediterranean, Princesse-Alice and the Prince undertook nearly all the kinds of studies familiar on modern oceanographic cruises. But the peak of Albert's career at sea was reached with a new vessel, Princesse-Alice II, which was in service between 1898 and 1910. This beautiful 1,420 ton vessel spent its first season at Spitsbergen with the Prince, the Scottish Antarctic explorer W.S. Bruce, John Young Buchanan, the chemist from Challenger, and the Kiel zoologist Karl Brandt; it returned to the Arctic in 1899, 1906, and 1907.

Princesse-Alice II, staffed by the Prince, Jules Richard (1863–1945), and a number of visiting scientists, was used for a wide range of scientific studies, including deep-water dredging and trawling, work on the midwater fauna, and studies of oceanic bacteria. Her deep trawling at 6,035 meters southwest of the Cape Verde Islands in August 1901 was the deepest successful bottom sampling in the Atlantic until the late 1940s. Hand in hand with this ambitious sampling program went the development, testing, and modification of new equipment for sampling throughout the depths of the oceans. As Prince Albert stated, "on exploite tous les systèmes imaginables pour étudier la faune des mers" ("one exploits every system imaginable in order to study the fauna of the seas"). By the early 1900s, he was devoting equipment and efforts to a few major projects, notably detailed studies of the vertical ranges of oceanic animals, and in 1909 he and his crew and scientific staff worked for five days west of Portugal in 5,940 meters of water doing the most complete oceanographic study in early oceanography.

Prince Albert planned for even more extensive studies of the open oceans after 1910, replacing the worn-out Princesse-Alice II with the fine steam yacht Hirondelle II, 1,600 tons and eighty-nine meters long, which began work in 1911. Concentrating on deep-water plankton collected with a variety of gear, including huge, rapidly-towed nets, the Prince and his associates studied the distribution and vertical migrations of midwater plankton. But this period was brief; when war broke out in August 1914, the ship returned to Monaco, and, despite a short cruise in 1915, never resumed oceanographic work. After the war, financial difficulties and the Prince's increasing ill-health prevented the renewal of oceanographic cruises based in Monaco. Recognized as an outstanding figure in the new science of oceanography, honored by societies and universities throughout Europe and in the New World, Albert died in 1922, leaving his legend, a lengthy series of publications by his scientific co-workers, and the Musée Océanographique de Monaco as his memorials.

The origins of the Musée Océanographique de Monaco lie early in Prince Albert's career as a zoological collector. In 1885 he had begun a collaboration with Jules Malotau, Baron de Guerne (1855–1931), a French nobleman trained in zoology in Lille. After moving to Paris in 1884, de Guerne studied marine zoology at Concarneau and at Wimereux, both sites of marine biological stations. By 1886 he had been appointed to take charge of the zoological work on Hirondelle, then on Princesse-Alice, tasks which continued until 1894.

De Guerne soon came in contact with Jules Richard, a young zoologist from the Auvergne who had begun zoological studies on freshwater crustacea. Their interests in crustacea brought Richard and de Guerne together. Shortly thereafter, in September 1887, the young Richard, upon de Guerne's recommendation, was appointed scientific secretary to Prince Albert in Paris. The following year he went to sea on Hirondelle from June 24 through September 21, reporting enthusiastically to his family about the equipment of the ship, the personalities of his companions, and the novelty of zoological work on the open ocean. Scientifically competent, hard-working, reliable, and sturdy, the redoubtable Richard, who completed a doctorate in 1891 and a medical degree in 1900, soon rose in rank, becoming chief of Prince Albert's scientific secretariat in 1894. The die was cast: the rest of Richard's life was devoted to Prince Albert, his foundation of the Museum, and his ideals.

The idea of a zoological station in Monaco had occurred to Prince Albert as early as 1885. Europe was in the midst of an explosion of new marine stations: Ostend in 1843, Concarneau in 1859, Arcachon in 1867, Roscoff and Naples in 1872, Wimereux in 1874, Kristineberg in 1877, Villefranche and Banyuls-sur-Mer in the 1880s. The great problems of marine invertebrate phylogeny, morphology, and embryology became the subjects of study from the Baltic to the Mediterranean. Classical laboratory zoology was revivified amongst enthusiastic young scientists working on living animals under, admittedly, often very pleasant human living conditions. With the founding and development of the marine stations came the palpable excitement of new discoveries.

Prince Albert was tantalized by the revolution in zoology taking place around him, so that it is no surprise to find that he had plans drawn up for a marine station, to be located at Fontvieille, to the west of Monaco-Ville, made up of a public aquarium, laboratories of biology, chemistry, and physics, and a library, along with working space for visiting scientists of high qualifications, and of any nationality. But this plan was not realized. Instead, he began to develop the idea of an oceanographic museum. The stimulus was the *Exposition Universelle* of 1889, held in Paris, at which he displayed some of the collections from the cruises of *Hirondelle*. The public response seems to have convinced the Prince that his collections, housed until then in his crowded hotel rooms in Paris, should have a better home, one where his international scientific ideals could be achieved—a place of study and public education where the high aims of science, as he viewed them, could be made manifest to a European world badly in need of stable foundations.

The new Museum began slowly. During the 1890s, Prince Albert was occupied with his duties as Sovereign Prince (beginning in 1889), with the building and equipping of *Princesse-Alice* and her successor, with the production of his first current chart of the North Atlantic, and with increasingly unpleasant relations with de Guerne, who had been editing the lengthy series of monographs in which results of the cruises were published. But with Richard securely in place as the new editor of the *Résultats*, and with *Princesse-Alice* II well underway, Prince Albert commissioned plans and called for tenders for a *Musée Océanographique de Monaco*, to be built on the cliffs beside the Jardins Saint-Martin. On April 25, 1899, the cornerstone was laid and building was begun. The following year, in December 1900, Richard was appointed director of the new Museum, moving to Monaco a month later.

For the next ten years, the Museum was the center of Richard's life. By 1908 the problems of building on a cliff site had been solved and the shell of the Museum was finished. Arranging the interior, subject to frequent changes of plans, took longer, but by 1910, when the official inauguration took place, the laboratories, an auditorium, a library, and two of the display halls had been completed. The last public display hall opened in 1913, coinciding with an International Congress of Zoology held in Monaco.

The Museum that existed in 1910—or better in 1913—is still recognizably that which is visited by hundreds of thousands today. In 1910 the Prince described it metaphorically as, "like a ship anchored along the coast bearing riches drawn from the depths; I have given it to scientists of all nations as an arch of alliance." Concretely, the building rose straight from the foot of the cliffs, having—just above ground level toward Monaco—a large entrance hall with an exhibition hall and conference room in the wings, an upper floor with large exhibition halls in both wings, and—below ground level—floors devoted to laboratories, administrative offices, a library, and the aquarium. The display halls then and now contain some of Prince Albert's collection of deep-sea animals, displays of the uses to which marine materials may be put, physical oceanographic instruments and accounts (today) of tides, tidal power, currents, weather, seawater, the uses of satellites and other topics of physical oceanography, while the central areas contain mementoes of the Prince, models of his ships, a laboratory from *Hirondelle* II, and new, modern topical displays intended to catch the eye of visitors with examples of recent technology ranging from satellite images to small research submersibles from the French fleet. Externally, the Museum was meant to be grand, decorated with marble columns, statuary allegorically showing the triumph of scientific truth and progress over ignorance, and, on the parapet, the names of twenty important oceanographic research vessels selected by Prince Albert and Richard. Above all was a large roof terrace, extending the length of the building (now with a café), intended to house meteorological instruments.

For the Prince, who endowed the Museum with four million francs (followed by a further million upon his death in 1922), the stability and permanence of his Museum was essential. After complex negotiations involving, in part, Dr. Paul Regnard (1850–1927), of the *Académie de Médecine* in Paris, Albert agreed with the French government in 1906 to incorporate the Museum into an Oceanographic Institute, legally a French private foundation located in Paris. With Regnard as director, while Richard retained his position in Monaco, the Paris Institute was provided with a building in the university quarter and inaugurated in January 1911. Its duties were to be the teaching arm of the foundation, while the Museum was to conduct research and to house the Prince's collections and public displays. Three professorships in Paris, which still exist today, were established, the subjects being biological oceanography, physical oceanography, and the physiology of marine organisms. Both the Paris Institute

A letter from Prince Albert I to Jules Richard, congratulating Richard on the award of his doctorate in 1891 and advising him, for his professional security, to begin work on a medical degree. Archives of the *Musée Océanographique de Monaco*; reproduction by Y. Bérard

The *Musée Océanographique de Monaco*, inaugurated by Prince Albert I of Monaco in 1910, faces the deep water of the western Mediterranean Sea. Photo: C. Olivi

and the Museum were to be governed by a high-level *Conseil d'administration*, and, overseeing the scientific activities, a *Comité de perfectionnement* made up of thirty eminent French and international scientists—arrangements that still exist today. More than seven decades later, the Institute in Paris has the same role, modernized and expanded, of providing university-level lectures and public seminars. It still conducts research and in 1978 opened its *Centre de la mer et des eaux* (Center for the Sea and Fresh Water), to educate the public about environmental problems. In 1907, to aid its work, the Museum was given a steam launch, l'*Eider*, which was active until the outbreak of the First World War, and was replaced by another of the same name in 1922. Meteorological instruments were acquired beginning in 1908; by 1911 they were functioning on the roof and in the neighboring Jardins Saint-Martin. The Prince himself set an example for the teaching role of the Paris Institute by establishing a series of lectures in Paris, some of the first series being given by Professor Julien Thoulet (1843–1936) of Nancy in the *Conservatoire National des Arts et Métiers* in 1903 (the later series being given in the Sorbonne). Public lectures, as well as university teaching were then—and remain—a major feature of each year at the Institute, even though it still has no formal connection with a university.

The years between about 1908, when the building was nearly finished, and the First World War, were active and creative ones at the Museum. A good deal of research was undertaken, based on the use of l'*Eider*, including Alexander Nathansohn's important study of the control of the annual plankton cycle in the Mediterranean, and the laboratories were put to use by zoologists, botanists, physiologists, and others. A stream of French and foreign visitors came to learn oceanographic or chemical techniques, often for use on major expeditions, and to examine the equipment in use in laboratories and on l'*Eider*.

The results of the Prince's annual cruises appeared in a series of magnificent volumes, titled *Résultats des campagnes scientifiques accomplies sur son yacht par Albert Ier, Prince Souverain de Monaco* (*Results of the scientific studies conducted by Albert Ier, Sovereign Prince of Monaco, on his yacht*), beginning in 1889. By 1910, thirty-four volumes had been published, with de Guerne, then Richard, as editor; during the next twenty-two years a further fifty volumes were printed (the total is now 110). Each represented intense work by the individual authors, but especially by the tiny staff of the Museum during editing and production. To aid the publication of short papers, intended to be published quickly, the

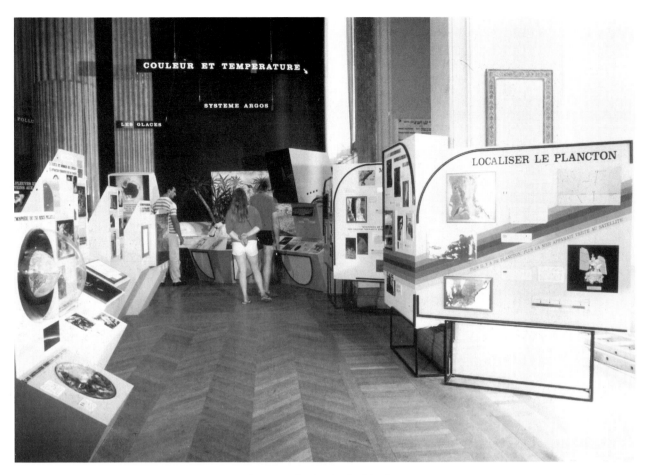

A recent display in the *Musée Océanographique de Monaco*, which presents the uses of modern technology, especially satellite-borne sensors, in determining sea temperatures, the abundance of marine animals, and the extent of sea ice. Photo: D. Noirot

Bulletin du Musée Océanographique de Monaco (renamed *Bulletin de l'Institut Océanographique* in 1907) was begun by Richard in 1904. On its part, the Institute in Paris initiated its own publication, the *Annales de l'Institut Océanographique* (modeled on the *Résultats*), in 1910. A steady stream of papers on marine biology, marine ecology, and physiology issued from the Museum and the Institute, accompanied occasionally by papers on oceanographic techniques, or lectures delivered at the Institute in Paris. Throughout, however, the most important subject was the scientific study of the Prince's rapidly growing collections, and after his death, the description of the specimens stored in the treasure trove of the Museum under Richard's supervision.

Among the most notable early successes of the Museum were its work on an authoritative bathymetric chart of the oceans, and its aquarium. Prince Albert had some experience with publishing charts, notably his North Atlantic current chart, when Julien Thoulet suggested in 1899 that he undertake the preparation of a bathymetric chart. Thoulet's suggestion was presented to the Seventh International Geographical Congress in Berlin the same year and adopted; a commission was established to oversee the production of the first General Bathymetric Chart of the Oceans (GEBCO) which appeared, thanks to the work of Prince Albert's staff, in 1905. A second edition followed beginning in 1912. By the 1920s the task was exceeding the capacity of the small staff in Monaco, especially after Prince Albert's death; as a result Richard suggested in 1928 that the work be taken over by the International Hydrographic Bureau, established in Monaco in 1921. The IHB has published all subsequent editions of GEBCO since the third edition of 1932.

The aquarium was one of the triumphs of the Museum. Richard took the first steps to set up an aquarium, intended for both research and public viewing, in 1901–02. The overworked Richard relinquished the task to a new employee in 1907. This was Dr. Mieczyslaw Oxner (1879–1944), with a doctorate from Zurich, who was hired as chemist and accounts secretary. The energetic Oxner took on the aquarium as a personal challenge, increasing its size regularly—it doubled in size in one year alone, 1924—experimenting with new displays, and introducing what he called "submarine landscapes," that is, ecological groupings of animals in their habitats, in 1925. The aquarium became, and remains, the great drawing card of the Museum, famed for the variety of living species it maintained and for its beauty. Oxner was particularly interested in the behavior of animals; he encouraged scientific observations of the living ani-

mals and recorded their behavior on film. Moreover, he was determined to display animals that were kept alive and well in captivity for long periods, representing real scientific knowledge of their living conditions and nutritional requirements. He also believed that public sympathy for marine animals could be fostered by teaching them to accept food from the hand, to be handled by keepers, or to learn simple tricks. As he said, in 1932, in describing his program at the aquarium, "the more one observes, the more one begins to understand life in all its infinity of manifestations."

The First World War, followed by Prince Albert's death, had profound effects on the Institute and the Museum. Both were funded solely by Prince Albert's original endowment, and later by a bequest. The war greatly reduced attendance at the Museum and postwar inflation bit deeply into its fixed income. Few researchers came to the Museum after the war, despite its fine facilities. An American visitor noted in 1926 that excellent laboratories and even furnished rooms for lodging in the building were available at reasonable rates, and that several months in the Museum, including an ocean passage from North America, would probably be less expensive than working for a few months in an American marine biological station, where costs were much higher.

But the great depression and other factors combined to keep the Museum struggling through the 1920s and into the 1930s. Richard appealed for visiting researchers in 1938, noting that he could place at their service bacteriological, physiological, and chemical laboratories, abundant working space in ten new laboratories below the aquarium, the use of the steam-launch l'*Eider* and motorboat *Pisa* in addition to a meteorological service and an excellent library. Grants, *bourses d'études*, were available (then as now) to increase the incentive to work in Monaco. But the depression, then the onset of the Second World War, combined to make the survival of Prince Albert's foundations in Paris and Monaco even more difficult.

During the Second World War the aging Richard could do no more than attempt to maintain the integrity of the Museum. During the Italian and German occupations he used his considerable authority to keep military units out of the building, thus saving it from direct attack during air raids and the Allied invasion in 1944. Research was impossible; attendance at the Museum was reduced to a trickle; the aquarium was neglected and the tropical fish died. Tragically, in 1944, Oxner, a Jew, was arrested by the Italians, then the Germans, and deported to Auschwitz, where he died. Richard himself did not survive the war. He died in

The first General Bathymetric Chart of the Oceans was published in Monaco in 1905 under the direction of Prince Albert's scientific secretariat. Above is the title page of the first edition, and below is a representative chart, from the second edition, showing part of the western Mediterranean Sea. The GEBCO, now coordinated by the International Hydrographic Bureau, is still closely linked with Monaco. Photo: A. Malaval

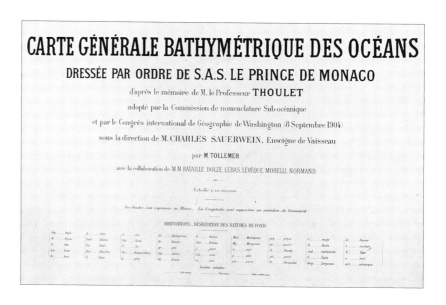

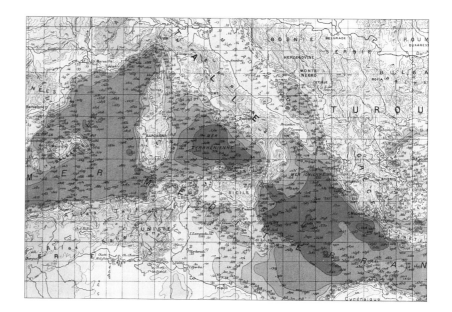

January 1945, while the liberation of Europe was in progress.

With the death of Richard the long period of direct connection with Prince Albert and his scientific ideals came to an end. Richard's successor, Commandant Jules Rouch (1884–1973), antarctic explorer and Professor of Physical Oceanography in the Institute in Paris, was faced with the task of reconstruction. The aquarium was rebuilt and restocked by Jean Garnaud; political and diplomatic links necessary to the prosperity of the Museum were restored and amplified, and a growing list of patrons subscribed to *Les Amis du Musée Océanographique de Monaco* (*The Friends of the Oceanographic Museum of Monaco*), a quarterly newsletter produced singlehandedly by the literate and prolific Rouch between 1947 and 1957. Attendance climbed; during the 1920s it had reached nearly 100,000 per year; during the Second World War the figure fell as low as 25,000. But during the late 1940s, and after, it rose steadily: about 170,000 in 1946; more than 380,000 in 1950; nearly 650,000 in 1957, the year of Rouch's retirement. Although scientific research was not one of Rouch's priorities, a slow stream of scientific visitors, mainly from Europe, came to work at the Museum, usually for short periods. Somehow, during the late 1950s, the Museum remained outside the sweep of reorganization and expansion in oceanography that was beginning then, despite the very practical suggestion by Louis Fage (1883–1964) to the *Comité de perfectionnement* (in 1950) that Monaco was especially well placed to be developed as a major center for deep-sea research because of its proximity to deep water. Swedes, Danes, and Americans took up the challenge of deep-sea research in the 1950s and 1960s, but Monaco did not. However, its aquarium had been rebuilt, exhibits had been renewed and expanded, and the crowds of public visitors appeared to grow without limit. Laboratories and services were available for scientists, even though a strong scientific program had not been developed.

The Museum took on a new role shortly after the accession of its new director, Commandant Jacques-Yves Cousteau (b. 1910) in 1957. Trained as a French naval officer, Cousteau had been a developer, with the engineer Emile Gagnan (1900–1984), of the aqualung. He was among the organizers of the French Navy's *Groupe d'études et de recherches sous-marines* (Underwater Research and Study Group), based at Toulon, which further developed SCUBA. Later, in the late 1950s, he was among the designers and planners of small sub-

mersibles, notably the first *soucoupe plongeante* (diving saucer), which became available to researchers in 1960. Cousteau's style was to leave day-to-day administration of the Museum to assistants, while he worked on diving equipment, spending more and more time at sea on his vessel *Calypso*, a converted American minesweeper. Cousteau's name, which has become synonymous with the Museum, with marine explorations, with the protection of the sea, and with a series of beautifully-produced films and books on underwater life, is now inseparable from that of oceanography in general, at least in the eyes of the worldwide public. His influence, combined with increasing mobility and affluence throughout the world, has made the Museum more popular—annual attendance in the 1960s, varying from just over 500,000 to 650,000, has risen steeply during the 1970s and 1980s to more than a million in 1984 and to a solid 900,000 or more since then, ranking it fourth as an attraction, after the Eiffel Tower, the Louvre, and Versailles. What the hundreds of thousands see is a building conceived by Prince Albert and Jules Richard, displays ranging from those prepared for the Prince by Richard, Oxner, and their small staffs, to recent French minisubmersibles and the latest museologically-sound displays of modern physical oceanography, along with the aquarium.

Behind the scenes, after several decades of reduced scientific activity, the Museum resumed its scientific work, partly through contracts directed to it through the *Comité Exploitation des Océans* (COMEXO), which funded projects as varied as studies of deep-water pelagic animals, the establishment of a radio navigation grid along the Côte d'Azur, and the building of a *bouée-laboratoire*, (buoy laboratory), a moored structure akin to the American FLIP, used for oceanographic work requiring a very stable platform. The Museum was also able to renovate and expand·its laboratories, adding one third to the space available for scientific research. But by the early 1970s, after COMEXO was transformed into the French government's oceanographic group, the *Centre National pour l'Exploitation des Océans* (CNEXO), contract funds dried up and the Museum again had to struggle to maintain its research activities.

Fortunately, at the time of the first COMEXO grants in 1960, Prince Rainier III had inaugurated the *Centre Scientifique de Monaco*, to be housed in the Museum. Its first operational unit was the *Laboratoire de radio-activité appliquée* (Laboratory for Applied Radioactivity), with the aim of measuring low-level radioactivity in the Mediterranean marine environment around Monaco, especially on particles in the air, in rain,

Various animals from the aquarium of the Museum, depicted by Paul A. Robert, showing tube worms, sea urchins, sea squirts (tunicates), and shrimps. Collections of the *Musée Océanographique de Monaco*; reproduction by M. Bourgeacq

seawater, sediments, and marine organisms. The Center expanded in 1966 with the opening of its *Laboratoire de microbiologie et d'étude des pollutions marines* (Laboratory for Microbiology and the Study of Marine Pollution), originally intended to monitor bacteria levels in Monegasque coastal waters, but now responsible for broader biological, chemical, and physical oceanography of the area in addition to the operation of the meteorological observatory. Even more recently the Center has incorporated the *Laboratoire de neurobiologie moléculaire*, (Laboratory for Molecular Neurobiology), specializing in basic neurophysiology and the effects of pollutants or poisons upon neurophysiological activity in invertebrate animals. Since its formation, the *Centre Scientifique de Monaco* has expanded its research—in the late 1980s to cover topics such as environmental monitoring and impact assessment, the distribution of heavy metals in the French and Monegasque Mediterranean, basic biological, chemical, and physical oceanography necessary for baseline studies, seismology, and meteorological analyses and statistics.

Only a year after the inauguration of the Center, the Museum, through the *Institut Océanographique*, entered into an agreement with the International Atomic Energy Agency (IAEA) and the government of Monaco to establish an International Laboratory of Marine Radioactivity (ILMR) to be housed in the new laboratories and funded by both the IAEA and Monaco. It rapidly grew in reputation, becoming noted for its studies of the transfer of radio-isotopes through marine food chains, its work on the transuranic elements such as plutonium, americium, and caesium in seawater, and most recently for studies of non-radioactive pollutants such as pesticides and metals. A very important function of this laboratory was its development of calibration techniques for analyses of radioactivity and its provision of quality-control analyses for many other laboratories. Early in 1988 the ILMR moved to a new laboratory on land provided by the government of Monaco in the newly expanded quarter, Fontvieille. In addition, the Museum built links with the Faculty of Sciences of the University of Nice to carry out work on marine ecology and physiology and used the Institute's twenty-meter research vessel *Winnaretta-Singer* both for projects at home and for fisheries studies around the Mediterranean coordinated by the *Commission internationale pour l'exploration scientifique de la mer Méditerranée* (CIESM), with which the Museum has been connected since it was founded under Prince Albert's patronage.

The *Musée Océanographique de Monaco* is probably the best known—certainly the best visited—oceano-

During the fiftieth anniversary
of the inauguration of the
Musée Océanographique de Monaco
in 1960, its director (1957–88),
Jacques-Yves Cousteau
(middle), was joined by (from
left to right), the Prince
héréditaire Albert, in the arms
of his father, Prince Rainier III,
Princess Grace, and Prince
Pierre (Prince Rainier's father).
They are viewing a portrait of
Prince Albert I. To the side are
water-collecting instruments
and thermometers from Prince
Albert I's time, and behind
them is a painting of the first
Hirondelle. Photo: G. Detaille;
reproduced with the kind
permission of the Archives,
Palais Princier, Monaco

graphic institution in the world. To a wide public it embodies the glamor and technical complexity of modern marine science and provides an opportunity to view living marine animals in beautiful settings. It has become the centerpiece of modern environmental awareness of the oceans, and a center for the study of the effects of man upon the oceans. But we have chosen to concentrate at least as much upon the history of the Museum (and the Institute of which it is a part) because of the unusual—indeed unique—role that it has played in the development of oceanography. In Monaco, to a greater extent and more vividly than anywhere else, we can still see and feel the history of oceanography. Because the Museum represents the past, the present, and perhaps the future of oceanography in a way not possible elsewhere, we believe that it has a function yet to be realized in presenting science to the wider world of affairs and scholarship.

Many of the public displays in the Museum still allow us to glimpse the past. But behind the scenes are thousands of scientific specimens, yet unstudied, now being fully catalogued, and a collection of scientific instruments unparalleled in its completeness. Prince Albert was fascinated by the invention of new instru-

ments of all kinds—water bottles, current meters, photometers, dredges, and trawls, to name only a few. Richard and Oxner continued the development and collection of instruments, several of them described recently in a splendid catalogue. But much remains to be done to complete the study of the instruments and to show how technology and science interacted to produce new instruments throughout the history of oceanography.

The contribution of the past to our present science of oceanography is also represented by the library and the written archives of the Museum. Prince Albert's and Richard's correspondence with most of the founding figures of late nineteenth and early twentieth century marine science is carefully preserved, but little has been studied. It presents a wonderful opportunity to link the present with the past, to study the ways that science grows or stagnates, flourishes or is held back, by the scientific and social forces under which it exists. These riches—instruments and archives—now receiving increasing attention from historians of science and technology, represent another facet of the future of the Museum, one based firmly upon its complex and fascinating past.

A radiocarbon dating laboratory of the *Centre Scientifique de Monaco*, located in the *Musée Océanographique*. Photo: A. Malaval

A part of the aquarium of the *Musée Océanographique*. Begun just after the turn of the century, it is one of the most beautiful of its kind. Photo: Y. Bérard

5.

The Alfred-Wegener Institute, Bremerhaven

Gotthilf Hempel

The Alfred-Wegener Institute
for Polar and Marine Research
in Bremerhaven. Main
building, opened in 1986.

Beginnings

Compared to the Portuguese, the British, or the Dutch, the Germans are not a people of great seafarers. Generally speaking, they have lived with their backs to the sea. It was only the great port cities on the coasts of the North Sea and the Baltic that developed efficient merchant fleets and ship-building industries. German oceanography had less of a basis in maritime traditions and interests to build on than oceanography in France or Britain. It was not until the end of the nineteenth century that national prestige and the requirements of the shipping and fishing industries offered oceanographers in Germany the political arguments to obtain funding for their expeditions.

The age of global oceanographic expeditions did not bypass German oceanography, which reached an early peak of success. This was initiated by the expedition of the *Gazelle* in the Atlantic, Indian, and West Pacific Oceans during the early 1870s. In South America, the *Gazelle* encountered the *Challenger*. The *Gazelle* expedition had five tasks—typical for that early phase of development of oceanography: geographic positioning, as the basis for all other work; regular meteorological observation; magnetic measurements; oceanographic investigations of currents and waves; deep-sea studies and soundings.

The most important subsequent expeditions were shorter and more specific in their tasks. Victor Hensen, sailing the *National* (1889), laid the foundation for the systematic investigation of oceanic plankton productivity in the open sea. In particular, he gave us a detailed description of the plankton zones in the Atlantic Ocean. Carl Chun led the *Valdivia* in 1898–99 on a deep-sea expedition to the South Atlantic and Indian Oceans that highlighted the contrast between the global uniformity of the deep-sea fauna and the zonal differentiation of the surface layers.

The research vessel *Meteor* crossed the Atlantic Ocean fourteen times in 1915–17, between 20° North and 64° South, measuring the vertical temperature and salinity distribution, and thus providing a general answer to the question of the circulation and stratification of the Atlantic water masses. The expedition also demonstrated how the zoning of phytoplankton and zooplankton, of nutrients and sediments is linked to the oceans' great current system and climatic zones.

That *Meteor* expedition is generally considered the most important single German contribution to oceanography. It initiated the phase of the systematic study of large oceanic spaces through high-precision measurements.

Side by side with this basic research in oceanography and the biology of the open oceans, a more specifically applied form of German oceanography began to emerge during the last quarter of the nineteenth century in the North Sea and the Baltic, catering to the needs and interests of the fishing industry. With the development of motorized trawling it became necessary to protect fish stocks from overfishing and to increase them, as much as possible, through artificial breeding and transplantation. This required research. Data were required, furthermore, concerning the hydrographic conditions of the German seas through changing seasons. For this purpose, a "Prussian Commission for the Exploration of the German Seas" was established, as well as a German Marine Observatory and a Biological Institute on the island of Helgoland. The formation of the International Council for the Exploration of the Sea (ICES) in 1902 responded to similar requirements, and was partly due to German initiatives.

Oceanography in a divided Germany recovered only slowly from the consequences of World War II. At first, it merely participated in the fisheries programs of ICES and in international cooperative projects in the North Sea and the Baltic. This was followed by investigations into the exchange between Arctic and North Atlantic water masses, and participation in the Polar Front Survey of the International Geophysical Year 1958. In this latter undertaking, two German research ships were involved for several months. In 1964 a new *Meteor* began its operations. This ship participated in the International Indian Ocean Expedition, which was a crowning achievement in the phase of multi-ship surveys.

In physical oceanography, as well as in marine biology, attention, at that time, was focused mainly on the question of seasonal and short-term changes in the marine environment. This required real-time simultaneous research over large areas, employing a large number of ships, following, as far as possible a

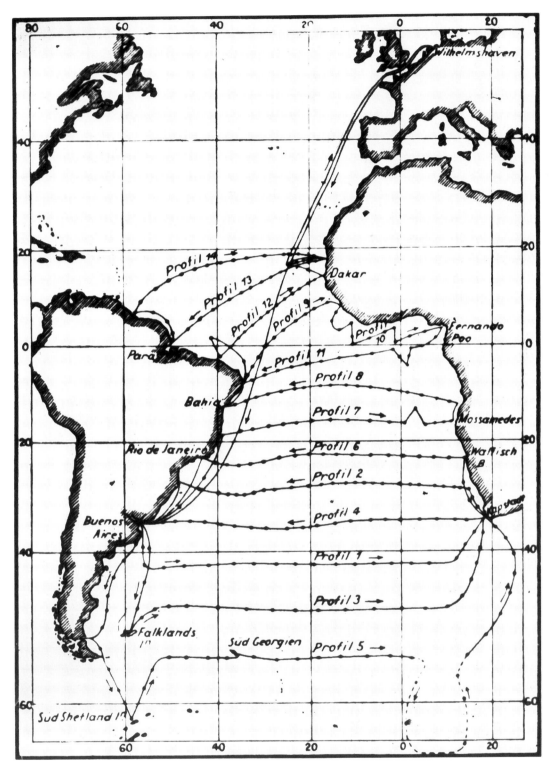

The German Atlantic
Expedition of R/V *Meteor*,
1925–27. Design: Alfred-
Wegener Institute

common, strictly determined routing and measurement program. Where this was not attainable, attempts were made to attract a large number of ships from many nations to an ocean area that had as yet been little explored, and to work on a common theme and with standardized methods of measurement, in order to obtain a common set of data covering all seasons. The Overflow Project on the Faroes island ridge is an example of the organized type of multi-ship survey; the Indian Ocean Expedition exemplifies the second, non-organized type.

After her return from the Indian Ocean, *Meteor* operated for twenty years in all parts of the Atlantic, from Spitzbergen to Antarctica, until she was sold to New Zealand in 1986. In Germany, she was replaced by a newly built ship which also bears the name of *Meteor*.

Structural Framework of Oceanography in Germany Today

Oceanography in Germany today has three specific characteristics: close cooperation between universities and governmental research institutions; international cooperation; and an interdisciplinary approach.

Oceanography in Germany is conducted by a multiplicity of institutions of different types and status. The Institute for Marine Science in Kiel is part of the University; smaller institutions are directly linked to the Universities of Bremen, Hamburg, and Oldenburg. The Max-Planck Gesellschaft supervises an Institute for Meteorology and is planning another one for deep-sea biology. The Alfred-Wegener Institute for Polar and Marine Research is a Foundation, funded almost exclusively by the Federal Ministry for Research and Technology which also funds, wholly or partly, the Helgoland Biological Institute and various other institutes and individual projects, while other Departments maintain special Federal institutions for applied sciences such as fisheries, geo-sciences, and hydrography. The scientists of all these institutions are cooperating closely. Common use of research vessels and special funding for cooperative programs through the German Research Association (*Deutsche Forschungsgemeinschaft*) contribute to the breaking down of institutional barriers. During the sixties and seventies, German oceanography developed rapidly—comprising today about 1,000 scientists and technicians.

Similar developments took place in the eastern part of Germany, the former German Democratic Republic. Most of the research there emanated from the

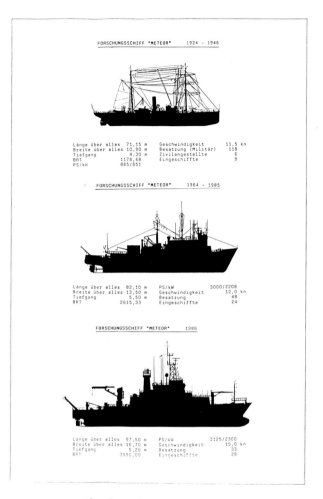

The three German research vessels, *Meteor* (from top to bottom): 1924, 1964, 1986. Reproduction: Alfred-Wegener Institute

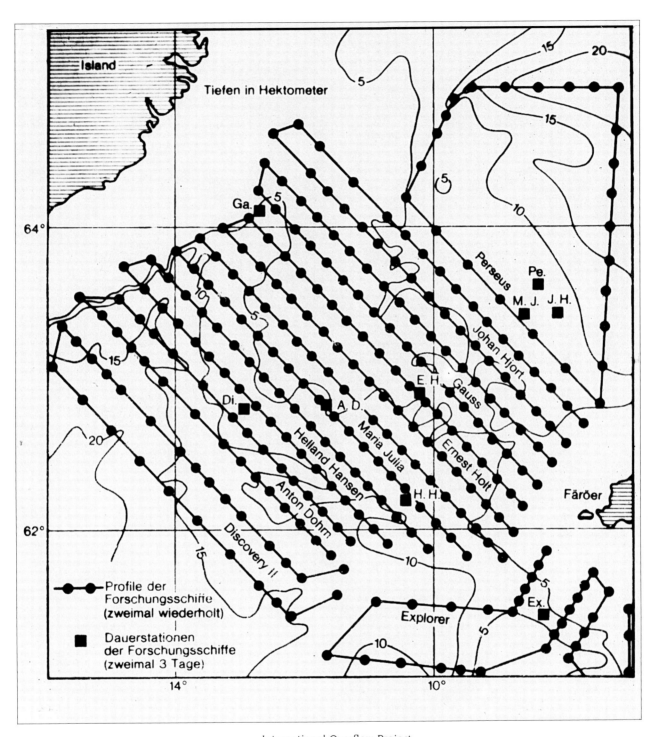

International Overflow Project
to study the water exchange
between the North Atlantic
and Norwegian Sea across the
Faroes ridge, 1973. Design:
Alfred-Wegener Institute

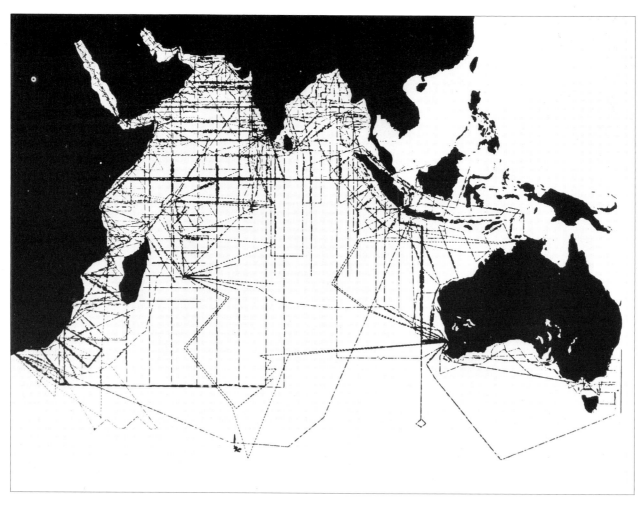

Cruise tracks of the
International Indian Ocean
Expedition, 1964–65. Design
Alfred-Wegener Institute

Institute of Marine Research of the Academy of Sciences in Rostock-Warnemünde, founded in 1958. Using two research vessels, this Institute concentrated upon investigations of the Baltic Sea, with special emphasis on water exchange and circulation processes, the pelagic ecosystem, bio/geochemical fluxes and budgets, and sedimentation processes. Additional research efforts concerning the dynamic processes of tropical waters and upwelling processes were undertaken in the Atlantic Ocean—mainly as contributions to various international programs. Marine research in nearshore areas was also carried out by smaller groups working within the framework of several universities.

International cooperation generally is a prerequisite for the conduct of oceanography; in a country like Germany, with its narrow access to the sea, it is of particular importance. A large number of young German scientists were exposed to international cooperation through scholarships giving them the opportunity to study in the United States or in England. Some of them remained abroad; most of them returned to Germany. They made an important contribution to putting an end to the isolation in which German oceanography had been stagnating due to pre-war and wartime conditions.

Most of the *Meteor* expeditions of the sixties and seventies were conducted within the context of international programs, coordinated by the Intergovernmental Oceanographic Commission (IOC), together with the World Meteorological Organization (WMO), or the Food and Agriculture Organization (FAO), as well as by the Scientific Committee on Oceanographic Research (SCOR), or by ICES.

The large programs of the United Nations organizations put great emphasis on cooperation with developing countries of the Third World to strengthen their oceanographic capacities. Scientists from developing countries participated in the *Meteor* expeditions, while quite a number of younger people from developing countries studied oceanography in Kiel and other German universities, or at various research institutes. German scientists also contributed to the building of institutes in the Third World. Occasional attempts to lend research vessels to Third-World institutions were less successful. Due to economic, technical, and human-resource conditions, these ships remained under-utilized most of the time.

Interdisciplinary cooperation is a natural given in German oceanography. The large Institutes in Kiel and Bremerhaven each comprise a broad spectrum of marine research activities, and in Hamburg attempts are made to integrate research in physical and chemi-

cal oceanography, marine biology, and marine geology. Even more effective have been cooperative multidisciplinary projects, funded over long periods of time by the German Research Association: e.g., the study of flora and sedimentation in Kiel Bay. In most cases, a *Meteor* expedition would comprise twenty-four scientists and technicians of various disciplines, and voyages of this kind naturally encourage a lively dialogue among the scientists, transcending the limits of separate disciplines—especially if the program itself has been planned in an interdisciplinary manner.

International programs were devoted to the interactions between ocean and atmosphere, such as the interactions between atmospheric and oceanic circulation during the monsoons in the Indian Ocean or the Equatorial Atlantic. Other studies were devoted to the effects of upwelling on ecosystems and fish stocks off the coasts of Northwest Africa or near the equator.

One should also mention some individual multidisciplinary projects in the Indian Ocean and in the Mediterranean, as well as some geo-scientific programs in the Pacific. However, the largest part of oceanographic research in Germany focuses on the North Sea, the Baltic, and the North Atlantic, from the Azores to the Polar Sea. Themes and emphases often depend on the interests of the leaders of the expedition. At the present time, the strong points in German oceanography are physical oceanography and marine meteorology, with special emphasis on the construction of large circulation models. Much work is being done on linkages between biology, geology, and chemistry in the study of the marine ecology, particularly in the context of marine pollution studies. Deep-sea exploration and the investigation of tropical seas are rather weak in German oceanography. The polar regions, on the contrary, are recently attracting a great deal of interest. We shall return to this later in this chapter.

Future Trends

Oceanography, all over the world, will have to focus both on regional and on global issues. Studies of the shallow seas adjacent to coastal States necessarily will concentrate on the question of sea uses (above all, fisheries, aquaculture, production of offshore oil and gas, sand and gravel, as well as tourism) and pollution of the marine environment; but coastal development, shipping, navigation, and defense will also play important roles. Every coastal State has its own interests, but there are problems transcending national boundaries, particularly with regard to fisheries and pollution, which call for regional cooperation. Many coastal States of the Third World, furthermore, depend on the assistance of industrial countries for the enhancement of their oceanographic capacity and the building of effective systems of monitoring and surveillance. For the oceanographers of Germany, the North Sea and the Baltic are traditionally the most important research areas. Baseline studies on the circulation and transfer of matter, ecological systems, and the nutrient cycle converge with investigations on the transfer and destination of noxious substances, the effects of eutrophication, and the development of fish stocks.

In the context of European cooperation, strategies are being developed to monitor and survey the North Sea. Large numerical circulation models, developed by the University of Hamburg, provide the most important part of the theoretical basis for this task. The Biological Institute of Helgoland, at the same time, has assembled the longest, uninterrupted series of observations on phytoplankton and nutrients in the German Bight, and the former Institute of Oceanography in Bremerhaven (now the Alfred-Wegener Institute) has, through various decades, made systematic observations on the demersal fauna of the southeastern North Sea region. For Kiel Bay one can make interesting comparisons between the data gathered on the demersal fauna during the late nineteenth century and today's data. Such long-term series of observations are essential to filter out long-term trends from the "noise" of short-term vascillations, and for a better understanding of the effects of occurrences such as storm surges, icy winters, or abnormal summerly periods of stagnation.

Problems relating to the world climate, summarized under the heading "Global Change," obviously are of deep concern to German oceanographers. As an industrial nation, with a high rate of energy consumption and important exports in the factory and machine industry, as well as in the chemical products sectors, Germany is a vigorous contributor to the "greenhouse effect," and it would be gravely affected by a sea level rise resulting from global warming. The global programs in this sphere offer an exciting challenge. Here is an opportunity to practically apply the experience in interdisciplinary, interinstitutional, and international cooperation that characterize German oceanography. What is at question is the role of the ocean as a heat engine and a sink of carbon dioxide and other greenhouse gases. In these processes, the Polar Seas are playing a particularly important role. Sea ice impacts on the heat and radiation balance and has a determining influence on the formation of deep-sea water, which absorbs large quantities of carbon dioxide from the upper layers of the sea.

Germany had joined the World Ocean Circulation Experiment (WOCE) with its research vessels *Meteor* and *Polarstern*, focusing on the ice-covered parts of the Greenland Sea and the Weddell Sea. The measurements taken from the ships will be supplemented with the continuously supplied data from anchored buoys that record important oceanographic parameters such as temperature, salinity, current velocity, and direction at various depths over a period of many years. They are coupled to sediment traps that collect sinking detritus, such as the remains of phytoplankton, the waste products of zooplankton, and inorganic dust. This process of sedimentation is an important part of the flow of carbon in the ocean. The Joint Global Ocean Flux Study (JGOFS) focuses on this problem area, with the purpose of establishing a relationship between primary plankton productivity and sedimentation in different regions and climatic zones of the world ocean.

In celebration of the centenary of Victor Hensen's Plankton Expedition in 1989, the *Meteor* joined an internationally manned expedition to the North Atlantic between the Azores and the Arctic. This expedition marked the official beginning of the JGOFS. At the same time, during the winter months, the *Polarstern* joined the Soviet research vessel *Akademik Fedorov* in the Weddell Sea to investigate the interaction between atmosphere, sea-ice and ocean water. The main problem to be investigated was the heat flow, the cloud formation, and the gas exchange within the pack ice zone. This expedition was planned by the Alfred-Wegener Institute jointly with the Soviet Arctic and Antarctic Institute in Leningrad. But American and British scientists participated as well.

Measurements at sea, carried out from ships and buoys have long since ceased to be the only source of data upon which German oceanography can rely. The development of global and regional climatic and ocean models is of enormous importance. With the help of ever more efficient computers, large ocean areas can be covered, with a vast spacial resolution and temporal extension. The hypotheses generated by this method concerning temperature and salinity changes, the movements of water masses and the resulting heat, are becoming increasingly more realistic. The time is fast approaching when it will be possible to predict the future effects of natural and anthropogenous influences on the oceans.

Another subject important for the future is the remote-sensing of oceans, sea-ice, and atmosphere

Weddell seals at their breeding
grounds on the fast ice at the foot of
the ice cliffs of the Riiser-Larsen Ice Shelf.
Photo: Alfred-Wegener Institute

from planes and satellites. Of particular importance for German oceanography will be the European earth-reconnaissance satellite ERS I, which began in 1990 to provide data on temperature distribution, sea level, waves, cloud and ice cover, among other things. Observation from satellites penetrates only the top layer—a few meters—of the oceans. Studies of the deeper levels depend on measurements from ships and buoys—until the day when, hopefully, there will be a breakthrough in remote-sensing technology.

None of the three great technologies—*in-situ* measurements, modeling, and remote sensing, can by itself solve the great problems of oceanography. It is the dialogue between field observation (including remote-sensing) and the modeling that will determine the future development of oceanography. It was with this assumption in mind, that the Alfred-Wegener Institute was founded.

The Alfred-Wegener Institute for Polar and Oceanographic Research

The Alfred-Wegener Institute in Bremerhaven is somewhat different from the other great European oceanographic institutions (e.g., Bergen, Kiel, Texel, Wormley), and obviously it cannot compete with the big centers in La Jolla, Moscow, Tokyo, and Woods Hole. The Institute is not homogeneous with regard either to age or function: The larger part was founded in 1980 as a Polar Institute; the older part, in 1919, as a Fisheries Institute. The small Fisheries Institute had developed, through the decades, into an internationally renowned institute for biological and chemical oceanography, with its main interest focused on the North Sea area adjacent to the German coast. With this orientation, it was natural that marine pollution became a primary issue as early as the sixties, even before it became a matter of public concern. A long-term series of investigations was initiated to monitor changes in the demersal fauna in the German Bight and relate them to the growing burden of noxious substances and nutrients. This research—systematically continued for decades—demonstrated a decrease in some particularly sensitive species, while others changed their habitat. Overall, however, it appears that the benthic biomass has increased rather than decreased, conceivably due to eutrophication from terrestrial nutrients. The consequences of fluctuations due to icy winters or periods of stagnation and oxygen depletion caused by summer heat, are far more dramatic.

On the one hand, the Institute became the home of specialists, employing world renowned taxonomists for diatoms, nematodes, actinia, and certain groups of microorganisms. On the other hand, it developed a high degree of interdisciplinary cooperation, especially in its North Sea investigations and the work on oil breakdown in the German *Wattenmeer* (mud flats). The phenomena generated on the front between the coastal waters and the open North Sea were jointly investigated by physicists, chemists, planktologists, and microbiologists.

For the sixty scientists and technicians of this well established and successful Institute, its incorporation, in 1986, into the new Polar Institute was rather disconcerting. Even though the word "Marine" was included in the Institute's new name, its main interest was to be concentrated in the polar regions. Over the years, however, the discrepancies between the older "oceanographers" and the newly arrived "polar investigators" began to erode. Quite a few of the "older" scientists chose the way from the North Sea to the Polar Seas; others continued their research in the North Sea. Taxonomists and chemical oceanographers, furthermore, are not restricted to any one region.

In the North Sea, new approaches could be developed within the context of European cooperation. The project EUROMAR with its international secretariat located at the Institute, promotes cooperation between industry and science on an international, European basis, in their search for new solutions to problems of oceanography and surveillance. The above-mentioned monitoring network for the German Bight was conceived as a pilot project. This network continuously monitors and records oceanographic and chemical data and transfers them to a central point where they are stored—but they are also instantly analyzed so that anything unusual can be immediately identified. This activity is linked to the development of automatic samplers and advanced numerical models of flow processes at sea. Another project is devoted to aerial remote-sensing with a laser-radar. All these projects are based on the principle that the scientists, who are the initiators of the projects, must find partners in industry for each project, and projects must not be limited to just one country. Industry and the institutes must bear part of the cost; the government pays the rest. These are the rules designed to enhance the development of advanced measurement instrumentation in Europe and foster cooperation between industry and science across national frontiers.

As the bad—sometimes exaggerated—news about the state of health of the North Sea accumulates,

Germany is trying to develop a research strategy which should provide the politicians with arguments for the adoption of effective measures to protect the North Sea ecosystem and restore ecological equilibria where they might have been upset. The demands for such arguments, coming from politicians and the public, are understandable, but, considering the complexity of ecological processes, scientists will be hard put to meet all of them.

Even the effects of fishing on the fish stocks often elude precise quantification; and it is practically impossible to neatly separate the slow, not immediately lethal, effects of noxious substances and nutrients from those of natural fluctuations. In the face of the impatience of politicians and public opinion, it is often hard for scientists to remain calm, to continue their patient search for the interactions within the warp and weft of the ecological system, to gather and analyze many long series of data, to simulate changes and transformations in their computer models—and yet, this is what is needed if we are to do more than issue generic warnings, if we would like to formulate reliable statements and predictions with regard to the possible consequences of the continued introduction of nutrients and noxious substances into the North Sea. Thus, on the basis of the already existing data, a measuring strategy is being designed, which should advance ecological research in the long term, and at the same time facilitate the immediate tasks of monitoring and surveillance of the North Sea. This is the task that has been assigned to the North Sea specialists of the Alfred-Wegener Institute.

The Institute had been established to provide at last a permanent home and fresh impetus to German polar research. When Germany lost its free access to distant fishing grounds as a consequence of the new Law of the Sea, and the Club of Rome made its dour predictions about the imminent exhaustion of important mineral resources, German politicians discovered an interest in the Antarctic. If Germany wanted to join the Antarctic Treaty regime, this would require a continuous contribution to scientific research in Antarctica. In order to be able to make such a contribution, German scientists demanded a special Institute from the Government. Attempts to develop a German polar research program had been made sporadically for more than a hundred years. Some individual expeditions were crowned with success, but after each expedition, ship and equipment had to be sold, because there was no Institute to assure continuity. Hence the German Parliament decided, in December 1979, to construct a research station on the Antarctic continent,

as well as to provide a research vessel and a supply ship. At the same time, it decided to establish the Alfred-Wegener Institute for Polar Research in Bremerhaven. The Institute was assigned three tasks: it was to conduct research itself; it was to provide logistical support for the polar research of other German institutes; and it was to coordinate German polar research at the national level, as well as assure international cooperation. As a National Science Center, the Institute is relatively independent; ninety percent of its funding comes from the Federal Ministry for Research and Technology. The remaining ten percent is contributed by the Land Bremen. Within its first eight years, the staff of the Institute reached 350 members, including the personnel of the former Institute for Oceanography. This does not include the crews of the Institute's ships and planes, because this personnel is provided by private firms that manage the ships and planes on commission from the Institute.

In contrast to most other countries engaged in Antarctic research, Germany put its main emphasis on marine scientific research. First of all, its scientists expected particularly rewarding results in this sector and, secondly, Germany did not wish to enter into competition with the founding members of the Antarctic Treaty who had been maintaining numerous stations on the Antarctic continent ever since the International Geophysical Year (1958). The maintenance of these stations, with the help of lavish ship and air logistical support, often consumes more than ninety percent of a research budget, so that very little money remains for science as such. Focusing instead on marine scientific research, Germany hoped for somewhat more proportionate results.

The construction of the icebreaker and research vessel *Polarstern* was completed in December 1982, when it was delivered to the Institute. 118 meters in length and with a displacement of 16,660 tons, *Polarstern* is the largest German research vessel and, at this time, the most versatile research icebreaker in the world. Accordingly, she is in very high demand internationally. During the first six years of her operation, almost 400 foreign scientists worked on board.

The most important areas of research of the *Polarstern* are the ice-covered zones of the Weddell and Greenland Seas. The ship is capable of cutting through sea ice up to a thickness of one meter, and can dispose of even much thicker ice through ramming. Thus science has acquired the possibility of rather free-handed research in the pack-ice zone. Recently, *Polarstern* has operated twice in the Weddell Sea, even during the winter months.

Artist's view of the future
EUROMAR monitoring of the
health of the German Bight of
the North Sea. Alfred-Wegener
Institute

The focus of attention during these pack-ice investigations is on the interactions between air, water, and ice. These have a powerful influence on the heat balance of the oceans and the atmosphere. Sea-ice is continuously moving, continuously opening up cracks from which large amounts of heat and moisture pass freely from the water (at −1°C) to the atmosphere (at −20°C). Up to 1,000 watts per square meter can flow from the cooling-down water into the atmosphere. As the seawater freezes, the salinity and density of the adjacent water rise. This process contributes, to a large extent, to vertical circulation. Thus the Weddell and Greenland Seas are the two "breathing holes" of the world ocean. It is there that surface water passes, through various levels, into the deep bottom layers, taking along oxygen and trace elements in the process. The role of the Polar Seas as a sink for carbon dioxide and other greenhouse gases is, at this time, a "hot subject" among climatologists. During its research, the *Polarstern* Expedition also studied the effects of the polar ozone holes and of the resulting increase in ultraviolet radiation on the ecosystems of the surface layers of the sea. This included measurements of plankton production and, consequently, the CO_2 balance—another topical polar research problem.

The formation of sea-ice from pancake and platelet ice can be ascertained from ice probes analyzed under a polarization microscope. In order to obtain such probes, teams are put onto the often kilometer-wide ice floes, while other groups take hydrographical measurements or plankton samples in the leads of open water.

The sediments in the Polar Seas contain the climatic record of past millennia. They hold the key to the reconstruction of the sea currents and ice ages of the past. Paleo-oceanography plays a big role in Germany as a bridge between oceanography and geology. At the universities of Kiel and Bremen, strong research teams are working on this, and the subject is also being investigated at the Alfred-Wegener Institute.

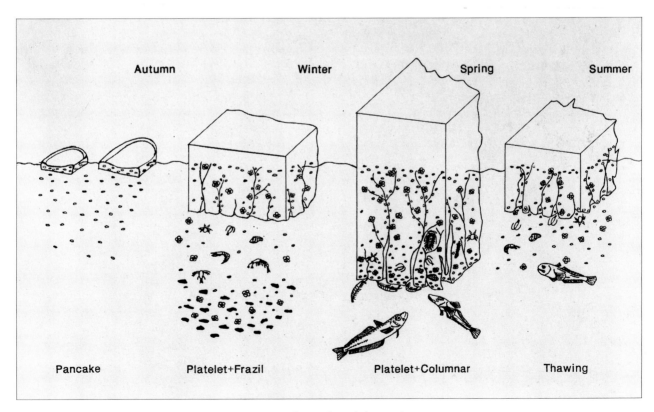

Seasonal growth and decay of
sea-ice and its communities
in the Southern Ocean. Photo:
Spindler

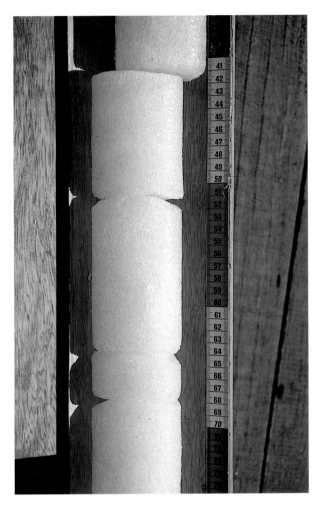

Drilling of ice cores in
Antarctic sea-ice (left), and an
ice core (above). Photo: G.
Dieckmann

The Institute's geophysicists are investigating the Weddell Sea's lower stratum and its shelf-ice. Both planes and ships are employed in this work. The planes are also used for the collection of meteorological data. During the past twenty years, remote-sensing from planes and satellites has become the most important instrument for extensive sea-ice research.

During the last few years, the Institute has made great strides in the modeling of the Southern Ocean and its connection with global circulation—especially since its research team was given its own computer center and direct access to the powerful mainframe computers in Stuttgart and Hamburg. A three-dimensional model of ocean circulation and of the ice covering of the Weddell Sea was developed with this help.

The European Polarstern Study (EPOS) serves as an example of the marine biological research activities of the Alfred-Wegener Institute and, at the same time, of a new form of international cooperation in marine research. In the period from October 1988 to March 1989, 130 scientists from eleven European and South American countries took part in this project. The expedition, which was divided into three sections, aimed at the investigation of the Weddell Sea's ecosystem. The European Science Foundation (ESF), in the name of the Alfred-Wegener Institute, invited West European marine researchers to submit research suggestions. More than 200 scientists replied. An international planning group developed a coherent program from their suggestions, and subsequently selected those who were to take part. About half of them had had experience in the Antarctic, the others wanted to apply ideas and methods which they had developed in other areas. The opening up of Antarctic research to scientists from countries and institutes which, until then, had no access to the Southern Ocean was an important aim of EPOS. Experience from West Greenland, the Barents Sea, and the coastal waters of Indonesia and Peru could be compared with that gained in the Weddell Sea: e.g., on the formation of pack-ice. The individual groups of scientists (e.g., for oceanography, phytoplankton, or fish) were multinational in composition in order to maximize scientific exchange. The evaluation of data and probes is also being carried out jointly. For this purpose, research posts have been made available at the Alfred-Wegener Institute and at various institutes abroad. The European Science Foundation has scheduled international seminars for evaluation purposes.

The sea-ice biosphere and the conditions for primary production in the marginal ice zone were the

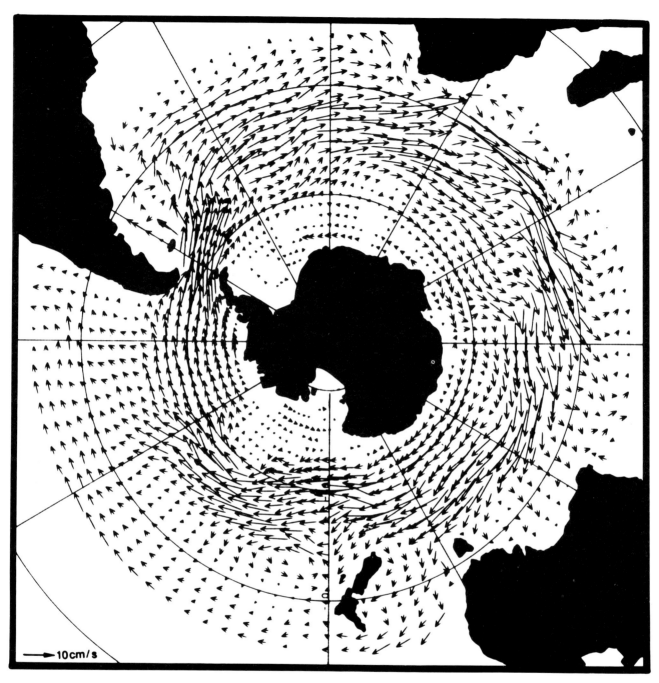

Velocity field of fifty meter
depth obtained from a high
resolution circulation model
of the Southern Ocean. Photo:
Olbers

topics of the first two sections of the expedition. Attached to the growing sea-ice, a highly diversified community of algae, unicellular animals and various small crustaceans emerges each year. They inhabit the underside, fissures and brine channels of ice floes. The water column under the ice is, in comparison, quite bereft of organisms. The water here is clearer than in any other part of the ocean.

It was at one time thought that under the sea-ice the production of plant life was restricted to the few ice-free months of the year. The fact that this assumption was false was shown by a voyage of the *Polarstern* deep into the pack-ice in the late winter of 1986. An underwater video camera supplied surprising evidence: the underside was teeming with krill. These approximately five centimeter long animals feed on algae that grow in and on the ice, also in the low winter sun.

Algal growth was more precisely examined with underwater video cameras and by divers on the EPOS Expedition. Algae grow most densely in deeply fissured ice. Such ice floes appear to have the effect of a light vault that catches the sunlight, guides it downward, and provides more light than is normally possible under flat ice. Rich occurrences of krill were observed by the researchers under algae-rich floes at the edge of the pack-ice. The krill scraped the algae off the ice.

Different types of planktonic algae from the previous summer were also frozen onto the ice floes. The melting ice set this plankton "seed" free, and the algae reproduced themselves in the water. A rich growth of plankton resulted directly in front of the retreating ice. At the same time, krill were freed from the melting ice and formed swarms between 30 and 100 meters below the surface. Krill is the most important consumer of siliceous algae off the ice edge. This was demonstrated on the *Polarstern* during the second section of the expedition. Two measurements, only two hours apart, revealed an astonishing reduction in algae concentration. A swarm of krill had consumed the algae underneath the ship. Later on, no more siliceous algae grew in this stretch of water. They were replaced by the smaller, short-lived forms typical of the summer.

On the third section of the expedition, the scientists devoted themselves to the rich supplies of fish and invertebrates on the seafloor, which are patchily distributed, and the species composition, which varies regionally. Catches by nets and grab samplers, as well as photos and videos of the seafloor, showed, above all, sponges, sea-lilies, sea cucumbers, and bryozoa. They feed on plankton remains which fall like rain onto the seafloor. The animals mainly cling to the bottom and feed from the near-bottom water layer. Their stiff skeletons are made of calcarious, siliceous, or horny substances. This protects them from predators, and they live to a ripe old age.

The rich Antarctic fauna also includes such long-lived vertebrates as whales, seals, birds, and fish. Almost all of them feed on krill, which can reach an age of five years.

Animal and plant life in the Antarctic is naturally adapted to very stable conditions, and will react sensitively to outside disturbances. Consequently, the EPOS Expedition also addressed the question of the ocean's ecological resilience in the hitherto little researched pack-ice zone of the High Antarctic.

The Alfred-Wegener Institute for Polar and Oceanographic Research in Bremerhaven was named after the meteorologist and geophysicist Alfred Wegener (1880–1930). He is most famous for his little book, *The Origin of Continents and Oceans*, which was published in 1915. While there had been earlier vague ideas about the drifting of the continents in the course of geophysical history, it was only Wegener who was able to document with geophysical and biological arguments his hypothesis of the successive break-up of a primeval continent, Pangaea, and of the drifting of the break-away pieces. In spite of that, the hypothesis was vehemently rejected by most contemporary geophysicists. Biologists, on the other hand, eagerly embraced it, since it furnished key arguments for bio-geography and the theory of evolution. The application of paleomagnetism and the deep-sea drilling of the *Glomar Challenger* were needed to prove beyond doubt the expansion of the seafloor starting from the mid-oceanic ridges, which gave rise to the Indian, and, later, to the Atlantic Ocean.

Working together with his father-in-law, Koeppen, Wegener made an important contribution to the development of paleoclimatology. Basing their work on the theory of continental drift, the two men published *The Climates of Past Geological Periods* in 1924. They were the first to link their explanation of the origin of ice ages to the long-term oscillations in solar radiation postulated by Milankovic.

Alfred Wegener was not only a great theoretician, but a most experienced practical meteorologist. At twenty-six years of age, he participated in a two-year Danish expedition to northeast Greenland. During a second expedition, with the Danish explorer J. P. Koch, in 1912–13, he traversed the Greenland ice sheet from east to west at 75° North. A third expedition, in 1930–31, was to explore the cross section of this ice sheet

Emperor penguins with their
chicks. Photo: Plotz

Alfred Wegener, geophysicist,
meteorologist, and polar
explorer, 1886–1930. Photo:
Alfred-Wegener Institute

utilizing a new explosive seismic process. At the same time, this expedition was to investigate meteorological conditions and changes throughout the year. The goal was to enhance the safety of transatlantic sea and air traffic. In November 1930, Wegener died from exhaustion during the return trip, on foot, from the station at the top of the ice cap. He was fifty years old.

One of the chief arguments made in favor of the Greenland expedition—which was extremely difficult to finance due to the global economic depression at that time—was that it offered an opportunity to train a new generation of scientists, and thus to ensure that Germany would not have to quit the scientific race of the international community in the field of polar research. Fifty years later, Wegener's wish was fulfilled, with the establishment of the Institute that bears his name. Wegener was not an oceanographer in the narrow sense, but indirectly he has stimulated oceanographers to think about the changing shapes of the oceans, their currents, the distribution of temperatures, and the living forms they have nourished in the course of the history of our planet.

Krill (*Euphausia superba*) is, by biomass, the most abundant crustacean in the Antarctic. In winter it stays in the pack-ice, while in summer it thrives in the clouds of phytoplankton at the ice edge. Photo: Alfred-Wegener Institute

6.

The P.P. Shirshov Institute of Oceanology, Moscow

A.S. Monin

Many diverse forms of life, such as tube worms and sea anemones, flourish in the deepest, coldest parts of the ocean. Some of them cluster together, resembling a bunch of flowers. Photo: Shirshov Institute

Introduction

The building on 23 Krasikov Street in Moscow is not large by modern standards, but it is quite spacious. Lined with glass, concrete, and white stone, it houses a well-designed and decorated interior with miles of stairs and corridors. Scores of doors lead into small studies and big labs crammed with equipment. They lead into storage rooms, where visitors may easily lose themselves among hundreds of racks and thousands of cans and boxes. Some of them open into the computing center, where there are young people, seemingly not busy at all, as if enchanted by the drone of the computers and the humming of air-conditioners.

Working in this building are the scientists of the Shirshov Institute of Oceanology of the U.S.S.R. Academy of Sciences. This is where they come, to be more exact, when they are in Moscow, not thousands of miles away at sea. They travel far and wide on board their research vessels, plying the waters of all the oceans and seas between the Arctic and the Antarctic. The point is that it is very difficult to make an appointment with an oceanologist. They are away from Moscow for months, on the high seas, or in Leningrad, Kaliningrad, and Gelendjik, where the Institute has its departments, or in aircraft taking them to the nation's seaports. The Institute of Oceanology was set up a mere thirty-six years ago. A historian will tell you this is nothing compared to the time scale with which he is dealing. A historian of science will, on the other hand, tell you that the Institute was formed in the long bygone days when only a handful of people on planet Earth could even imagine the huge scale of the ocean exploration in the immediate future—i.e., in our own lifetime—or the mind-boggling scale this kind of research would assume toward the turn of the century.

The ocean was man's pathway to all great geographical discoveries. Seafarers set keel to water and sail to wind long before the collective reason of many later generations brought forth the elegant and powerful ships of this day and age. Thor Heyerdahl's daring voyages on balsa rafts and boats of ancient design brought him world fame and reminded us all of the courage and seamanship of the seafarers of old, the Phoenicians, Greeks, Polynesians, and Normans, to name only a few.

After traveling the breadth of the North Atlantic on board a modern trawler, or covering the length of the Northeastern Passage in the Arctic on board a nuclear-powered icebreaker, you begin to marvel at the daring of the men of Leif Eriksson's crew who discovered America five centuries before Columbus, and the courage of the men of Semyon Dezhnev who voyaged all the way from the Kolyma river-mouth to the Pacific in small boats to discover the strait between Asia and America in 1648.

Man was driven by a mixture of curiosity and commercial instinct when he sailed toward the unknown lands over the horizon. The sea brought the Argonauts to the land of the Kolchis, where mountain streams carried so much gold sand that a sheepskin dropped into their waters became instantly covered with a fleece of gold. The seafaring folk of old traveled across the ocean for spices, slaves, and silver. They discovered and conquered whole continents. In the end all there was to discover was discovered, and the time had come to discover the ocean itself. The explorers of the ocean today command a degree of courage to match the courage of Columbus, Dezhnev, Cook, and Kruzenstern. They have far greater knowledge and elaborate theories of naval architecture, fleets of powerful surface ships, and underwater manned and unmanned vehicles, as well as satellite navigation and communication systems, sophisticated drilling gear, and the ultimate in television and photo instrumentation—but the challenge they face has not diminished. Indeed their task has become much more difficult. The ocean must be put to the service of the human race. That task is as immense as the ocean itself.

The risks the divers in manned submersibles are running are no less than those facing cosmonauts manning orbital labs. Far fewer people have descended to a depth of 5,000 meters than have gone up to an altitude of 200 kilometers. It is easier to assemble a space lab of units, all but factory produced on the ground, than to build an underwater habitat on the slope of the continental shelf. It is easier to protect oneself from space vacuum and temperatures in the neighborhood of absolute zero ($-273°C$), than to find reliable protection against high pressures and the dreaded "bends" caused by working in compressed air

A dish of caviar. Caviar can be produced from farmed and hybridized sturgeon. Surgical technology has been developed for extracting the caviar without killing the fish.
Photo: Shirshov Institute

in submerged caissons. The road to the treasures and mysteries of the ocean has never been an easy one. Scientists have succeeded in solving some of them, but there are a great many more as yet unresolved.

The Shirshov Institute of Oceanology: Origin and Evolution

The Shirshov Institute closely cooperates with all the leading research institutions of the U.S.S.R. Academy of Sciences dealing with mechanics, physics, chemistry, biology, geology, and geophysics. The Institute's extensive international scientific ties are rooted in the memorable events of the International Geophysical year (1957–1958). International cooperation in the studies of the ocean proved to be so fruitful that in 1960, the Academy of Sciences initiated the establishment of the Intergovernmental Oceanographic Commission of UNESCO (IOC). Before that, the Academy of Sciences became actively involved in the Scientific Committee on Oceanic Research (SCOR) formed in 1957 as a nongovernmental scientific body operating within the framework of the International Council of Scientific Unions (ICSU).

Members of the Institute's staff work in the specialized groups of SCOR. They are also involved in research coordinated by the International Association of the Physical Sciences of the Ocean. As soon as IOC was established, the scientists and research vessels of the Institute of Oceanology joined the International Expedition in the Indian Ocean (1961–66), and later they took part in the international studies of the Tropical Atlantic (1963–64), and all other international expeditions undertaken by IOC.

On March 10, 1921, V. Lenin signed a decree of the Soviet of People's Commissars on the formation of the Seaborne Marine Research Institute Plavmornin and the organization of its first expeditions based on the icebreaker ship *Malygin* and the first Soviet research vessel *Persey*, built in 1922. The decree elevated marine research to a level of important government affairs and called for every effort in its support at a time when the country was getting back on its feet after the devastation inflicted by World War I and the Civil War, and was still in the grip of widespread hunger.

The school of the *Persey* graduated remarkable researchers, who brought the Soviet school of science world acclaim. They continued the traditions evolved by Russian seamen and explorers, I. Kruzenstern, F. Belinghausen, of Antarctic fame, M. Lazarev, and the founders of the national school of oceanography, S.

Makarov, Yu. Shockalsky, and K. Deruygin whose works were distinguished by an in-depth approach to the processes and phenomena observed in the oceans. The approach adopted by the scientists of that school to the exploration of the ocean comprised the entire range of related sciences and gave a fresh and powerful impetus to the development of the physics, chemistry, biology, and geology of the sea. The fundamental studies carried out in the 1920s, over a relatively short period, laid down the scientific groundwork for further research efforts in applied oceanology, the charting of the Northeastern Passage, hydrography, and hydrometeorology.

The U.S.S.R. Academy of Sciences first showed interest in marine research in connection with the Second international Polar Year (1932–33), during which large-scale studies were undertaken on the initiative of the Soviet Union in the Soviet sector of the Arctic. At the time, and later, the outstanding scientist, geophysicist, and geographer, Academician O. Schmidt, organized many important scientific expeditions. Among them was the first drifting scientific station *North Pole* (*Severny Polyus*-1, 1937–38) and the heroic voyage of the icebreaker *Georgy Sedov* in 1937–40. The scientific information acquired was so varied and important that special coordinating measures had to be undertaken to comprehensively analyze it. In November 1939, the Presidium of the Academy of Sciences established its Oceanographic Commission headed by Academician P.P. Shirshov, who had taken part in the *North Pole*-1 expedition. In 1941 the Laboratory of Oceanology was set up by the Academy of Sciences, at first on a relatively modest scale. During World War II the Laboratory was evacuated from Moscow and transferred to Krasnoyarsk, but it continued its research work without a day's interruption. It was concentrating on in-depth theoretical studies. In 1943, Professor Shtockman joined the Laboratory. At the time he was largely interested in the dynamics of sea currents and the mixing of water in the ocean. His ideas were far ahead of contemporary thinking on dynamic oceanology abroad. In the 1950s they provided the foundation for an authoritative Soviet school on sea dynamics.

In 1945, the year Nazi Germany was defeated, the Institute of Oceanology of the U.S.S.R. Academy of Sciences was established. On January 31, 1946, by decision of the Presidium of the Academy, the Laboratory of Oceanology was transformed into an Institute headed by Academician Shirshov and his deputy V. Bogorov, who was a Doctor of Science in Biology. After a thorough discussion, a prophetic choice of goal was made by the Academy between studies of the World Ocean as a whole and studies of the Soviet marginal seas. The decision fell in favor of the World Ocean.

In 1949 the Institute sent the research vessel *Vityaz* on her maiden voyage; later, the *Vityaz* became the flagship of the Soviet scientific fleet. A new page was opened in the studies of the ocean by the Academy of Sciences.

The deck and laboratories of the *Vityaz* became a professional school for many scientists who now make up the backbone of the Institute's staff. The first voyages of the *Vityaz* were devoted to detailed investigations of the Far Eastern seas, and among them, the Sea of Japan, the Okhotsk, and the Bering Seas. Eight years later, in 1957, the *Vityaz* set out on her twenty-fifth mission and her first mission into the open ocean. That expedition to the Pacific initiated the list of major discoveries by Soviet explorers. One of these was the discovery of a developed and diverse life in the deepest parts of the ocean—at depths from six to eleven kilometers, in an environment of high pressures (600–1,100 times greater than at the surface), in total darkness and nearly freezing temperatures—where, until then, life was deemed to be impossible. As it turned out, life proved to be adaptable far beyond human imagination. Professor A. Ivanov established that the pogonophore is an independent and hitherto unknown species of bottom-living organisms. He was awarded the Lenin Prize (in 1961) for his monograph about it. Pogonophores are tube worms, similar to those more recently discovered near the hydrothermal vents, but without the red crests of the latter. Pogonophores have the peculiarity that they are mouthless. Like the hot-vent creatures, they feed through a symbiotic relationship with bacteria.

At the same time, the *Vityaz* depression, the deepest spot in the World Ocean, was discovered in the Marianna trough's southern section (11,022 meters below sea level).

During her thirty years in commission, the *Vityaz* went on sixty-five voyages and covered over 750,000 miles across the seas and oceans. The scientific results of her voyages, for instance, the discovery and exploration of major deposits of ferro-manganese nodules on the ocean floor, signified an important contribution to world science by Soviet researchers.

The evolution of the approach adopted by the Institute for its expeditions is highly significant. It began with a descriptive and fact-finding stage and moved forward to a detailed analysis and exploration of the complicated nature of processes and phenomena in the ocean and on the ocean floor. This change from description and geography to analysis and quantitative physics, geology, and biology of the ocean is the reason for the Soviet preference for the word "oceanology," leaving the name "oceanography" to the descriptive geographical part of marine sciences.

Now the expeditions of the Institute are highly specialized and centered on experiments in the ocean. The link between observations and theory is gaining in importance as specialized and experimental expeditions advance further. That is why, while enhancing the capability of the research fleet, Soviet oceanologists do not neglect the need for a parallel expansion and reinforcement of theoretical and experimental efforts in all avenues of contemporary science concerned with the ocean.

Among the examples of such an approach is the *Polygon-70* experiment in hydrophysics carried out in the tropical Atlantic by several research centers led by the Institute, and the underwater expedition to explore the Red Sea Rift organized by the Institute in 1980. Expeditions of that kind have demonstrated the enormous potential of planning specialized experiments in the ocean on the basis of principles predicted by theory. That approach has brought very good results. Following the discovery of synoptic-scale oceanic eddies by Soviet scientists during the *Polygon-70* experiment, a major Soviet-American expedition was organized to explore their physics and dynamics (POLYMODE). We will return to this at the end of this chapter.

The evolution of the approach to organizing expeditions proceeded against the background of a continued growth of the scientific potential of the Institute, and the accumulation of more information about the ocean. That information had to be generalized and analyzed, and this dictated the need for expanding the fleet and adapting it to new scientific problems.

In the mid-1950s, when the scars inflicted on the Soviet Union by the war had not yet healed, V. Kort, a doctor in geography and later a Corresponding Member of the Academy of Sciences, became Director of the Institute. The Deputy Director of the Institute, N. Sysoyev, a technologist, did a great deal to modernize the equipment of the Institute. The Scientific Council then discussed the crucial methodological problems of investigating the ocean as an integral natural entity whose giant scale called for concerted efforts by researchers in different sciences. At the time, the key problems to be resolved in various sectors of research were formulated. As a result, an overall picture emerged of the interrelations between the ocean and

Drop-off in the Red Sea.
Photo: © 1991 Carl Roessler

the atmosphere, and of the interaction of life and the oceanic environment. The principles of geographical zoning in the distribution of life in the ocean were laid down, and the details of the geographical and geophysical "face" of the ocean were generalized.

The extensive, and at times quite heated, discussions held at the Institute at that time elaborated upon the principles suggested by the leading oceanologists, Zenkevich, Shtockman, Bogorov, Bruyevich, Dobrovolsky, Rass, Bezrukov, and others. At that stage, the Institute of Oceanology evolved a scientific identity of its own and considerably enhanced its standing as a research center. In 1965, A. Monin, who held doctorates both in Physics and Mathematics, became Director of the Institute and served in this capacity until 1987. In 1967 the Institute, now a major oceanological research center, was named after Shirshov.

The Institute set up its first experimental station in 1949 in Blue Bay near Gelendjik on the Black Sea. In 1956, the base was used by an expedition consisting of two small research vessels, the *Akademik Vavilov* and the *Akademik Shirshov*. That expedition laid the foundations for the currently widely used *Polygon* method for the investigation of currents and turbulence in the ocean. The Gelendjik station has become the Southern Branch of the Institute. In recent years, new methods for the exploration of the ocean with the help of submersibles have been developed there. This base is also the training ground for the Institute's diver-pioneers of the underwater exploration of the sea.

The Pacific Department of the Shirshov Institute of Oceanology was established in 1961 in Vladivostok, and in 1973 it became an independent Pacific Oceanological Institute of the Far Eastern Scientific Center of the U.S.S.R. Academy of Sciences.

In 1961 the Marine Hydrophysical Institute handed over to the Institute of Oceanology its Department in Kaliningrad. Now it is the Atlantic Department of the Shirshov Institute of Oceanology, and a base for large research vessels operating in the Atlantic and Indian Oceans. The scientists of the Atlantic Department carry out important studies in physical oceanology and marine meteorology, as well as develop new instruments and sophisticated computer methods for processing hydrophysical data. In 1966 the Shirshov Institute of Oceanology opened a Department in Leningrad. The Leningrad Department comprises laboratories for dynamic meteorology, optics of the atmosphere and the ocean, among others. Directed by D. Chalikov, Doctor of Science in Physics and Mathematics, it carries out large-scale numerical modeling of the ocean-atmosphere system. There are so many aspects now to the scientific research of the Institute that they will be described separately here.

Physics of the Ocean

A great many crucial notions of the physics of the ocean were first voiced by Zubov, Berezkin, Shuleykin, and Shtockman. Their work laid the scientific foundation upon which narrow and specialized avenues of exploration of the ocean within the framework of thermodynamics, hydrodynamics, acoustics, and optics began to take shape.

The school of sea dynamics that has emerged at the Institute of Oceanology has achieved international acclaim. Having centered its research effort on the fundamental problems of the dynamics of the general circulation of the ocean, it began to specialize, starting from the mid-1960s, in the investigation of sea turbulence, internal waves, the interaction of the ocean and the atmosphere, and—beginning in the mid-1970s—in the investigation of the micro-structure and the fine thermohaline structure of the waters of the ocean (the Institute established the wide occurrence of these structures almost everywhere in the ocean), oceanic fronts and eddies of a synoptic scale, and mesoscales. In 1980, A. Monin, K. Fyodorov, and V. Shevtsov were given credit for the discovery of the phenomenon of the thin-layer movement of oceanic waters. V. Stepanov made an important contribution to establishing empirical climatology of the ocean.

Physical oceanology today is distinguished above all by a continued specialization, rapid theoretical advancement, extensive use of numerical modeling assisted by fast computers, laboratory simulation of individual physical processes and phenomena, and the aerospace remote-sensing of the ocean. However, for progress in physical oceanology to occur, an entirely novel notion of the organization of physical measurements in the ocean, that has evolved recently on the basis of specialized expeditions, is far more important than the continued specialization mentioned above.

The new, so-called *Polygon* organization of measurements stems from the clear-cut objectives formulated for each expedition. A polygon here is understood as a preselected site in the ocean, where a package program of physical measurements planned in advance is carried out over a certain period to resolve a range of problems of particular interest. That kind of approach allows for changing the program of measurements during an expedition on the basis of an approximate analysis of the data being acquired.

The optics of the seas and hydro-acoustics, being naturally linked to physical oceanology, have been greatly influenced by the need to resolve a range of important practical problems. Professor K. Shifrin, and later Dr. B. Kelbalhanov, coordinated the Institute's research into the optical properties of seawater and the laws of light propagation in the ocean determined by them. The Atlantic Department has focused on novel research trends in optics and, in particular, in photoluminescence. The staff of the Krasnoyarsk Institute of Physics of the Siberian Branch of the Academy of Sciences is assisting the biologists of the Institute of Oceanology in the exploration of bioluminescence in the ocean, which opens up good prospects for resolving some of the physical and biological problems of oceanology.

The exploration of the distribution, scattering, and absorption of sound in an oceanic medium against the background of natural noises has attracted much attention from Soviet physicists and naval specialists. The main event in this field was the discovery of the phenomenon of long-distance sound propagation in the ocean through the so-called underwater sound channel formed by particular vertical distributions of temperature, salinity, and pressure. The phenomenon was noticed by Soviet naval specialists during World War II in 1942, while using depth bombs against German submarines. It was studied later during a special naval operation in the Japan Sea in 1946, and the discovery was published the same year by N. Seegachev in a naval bulletin. Simultaneously, the phenomenon was predicted by the famous American marine geophysicist M. Ewing and studied in 1944–45 by special American Navy expeditions with the first publication by Ewing *et al.*, in 1946, in a geological bulletin.

Since 1953, studies in the U.S.S.R. have been concentrated at the Acoustics Institute of the Academy of Sciences, where many practical applications have been devised. These studies received a powerful impetus in 1961 following the launching of two twin acoustic research vessels, the *Sergei Vavilov* and the *Pyotr Lebedev*. Later, due to the initiative of Academy President A. Alexandrov, a new acoustic unit was established in the Institute of Oceanology. Recently, two new acoustic ships have been built for the Institute.

Marine Biology

The outstanding Soviet biologist L.A. Zenkevich, who concentrated his research efforts on the deep-water fauna, supported the concept of the biosphere by approaching the animal kingdom of the ocean as an integral whole. His fundamental work, *The Biology of the U.S.S.R. Seas*, is the basic reference book of several generations of marine biologists. The concept of the biosphere raised the problem of the biological structure of the ocean and paved the way to combining the qualitative methods of marine biology with a zoo-geographical approach.

The entire World Ocean—from the Arctic regions to the equatorial latitudes, from the sun-suffused top layer to the deepest oceanic depressions impenetrable by light—is inhabited by all kinds of living creatures constituting an extremely complicated biological structure which remains largely a mystery. The marine biologists' chief objective is to decipher the laws governing life in the ocean, so as to determine its biological productivity and, ultimately, to learn to control this productivity. It may indeed be possible to increase the volume of the ocean's biological product output tens and even hundreds of times. This means that in the ocean man must do what he has done on land—to pass from gathering to efficient farming. Artificial raising of biological productivity in the high seas should begin with an increase in the "yield" of plankton, using the same methods as those employed by the field farmer, i.e., by applying fertilizers (say, in the form of floating and slowly-dissolving granules of phosphorus and nitrogen salts, wherever they are deficient). Increasing plankton productivity could be combined with breeding industrial fish that feed on it, such as the anchovy. The Institute's scientists believe such experiments could be staged in the mesoscale eddies or rings of the Gulf Stream and Kuroshio. The Institute has worked out the first mathematical models of open-sea plankton communities; it is creating a numerical model of the biological productivity of semi-enclosed water surfaces such as the Black and the Baltic Seas; there are plans to use a simulation model to study various ways of increasing biological productivity of the oceans employing numerical methods. Notably, this research aimed at mastering the biological productivity of the ocean has been pioneered by the Institute's scientists.

Professor V.G. Bogorov carried out in-depth studies of the plankton and the bioproductivity of the ocean as a whole, and trained a large team of specialists in plankton studies, who are now exploring the primary biological products of photosynthesis, the primary links of the food chain in the ocean, the ecological systems of so-called upwelling zones which are characterized by the raising of deep water to the surface, and other regions in the ocean. That research is now

Spawn at bottom with some grains of roe swimming in water. Photo: Shirshov Institute

Spawn collected by biologists for analysis. It may be possible to increase the volume of the ocean's biological product output tens or even hundreds of times. Photo: Shirshov Institute

coordinated by M. Vinogradov and it relies extensively on the numerical models of biological populations.

The staff of the Institute of Oceanology has made important contributions to ichthyology. The works of Professor T. Rass and Dr. N. Parin concentrate in particular on the fish resources of the seas washing Soviet territory, and ways of replenishing them by acclimatizing new commercial species.

Marine Geology and Geophysics

Rapid progress in marine geological and geophysical research began in this country during the early years of the Soviet government. The first notions of the seabed relief and types of bottom sediments (based largely on the Black Sea) were advanced by N. Androusov, A. Arkhangelsky, and N. Strakhov in the pre-war period. Bottom sediments were investigated in the Barents and Kara Seas by expeditions on the research vessel *Persey*. Shortly before the war, the Lomonosov and Gakkel ridges were discovered in the Arctic Ocean.

After the war, marine geophysical and geological research was largely advanced by regular expeditions based on the *Vityaz*. These expeditions discovered and explored many major bottom relief forms, and among them, underwater ridges, mountains, troughs, and fault zones. The data acquired by them made it possible to compile geomorphological and tectonic maps of the oceans. The expeditions established the thickness and other physical parameters of the mass of bottom sediments and the oceanic crust in various tectonic regions. They mapped the distribution of the geophysical fields in the Pacific and carried out important studies of the geology of the bottoms of the Caspian, Black, and Mediterranean Seas.

In recent years, P. Bezrukov and A. Lisitsyn—both Corresponding Members of the Academy of Sciences—and other scientists have studied the problems of sediment formation, the distribution of various types of sediments, the composition and structure of sediments, and the variation of sedimentation processes. Their objective was to establish the sequence of the formation and the original spatial disposition of ancient oceanic sediments, with the help of isotope and paleomagnetic dating and geochemical techniques. Their studies provided the basis for the contemporary notions of the ferro-manganese nodules in the ocean and the distribution of oil and gas in the Soviet marginal seas.

Geological investigations of the ocean floor have enabled the Institute to do paleo-reconstructions of

Packing of fish in a fish-factory ship. New technologies for post-harvest conservation, transporting, processing, and waste utilization are amplifying the oceans' contribution to human nutrition. Photo: Shirshov Institute

Marine biologists working in a laboratory. The staff of the Shirshov Institute has made important contributions to ichthyology, working in particular on the fish resources of the seas washing Soviet territory. Photo: Shirshov Institute

various stages of the Earth's geological history. The paleo-reconstructions made for various ages based on the theory of continental drift objectively demonstrated, from the viewpoint of lithology and geochemistry, the indisputable advantages of the postulates based on new global tectonics (or lithospheric plate tectonics).

About 15 million square kilometers of the World Ocean's total area of 232 million square kilometers are believed to be worth prospecting for oil and gas. The Institute has developed entirely new theories of predicting the presence of oil and gas in various parts of the World Ocean floor; these theories take into account all the chief natural factors which give rise to and stimulate the processes of oil and gas formation under the ocean floor.

Special mention should be made of the Institute's experiments in extracting rare and scattered micro-elements from the Black Sea.

The Chemistry of the Ocean

The ocean can be likened to an enormous chemical reactor sustaining continuous complex chemical processes: ion interaction in a solution, ion exchange, gas absorption and emission, interface sorption, colloid formation and degradation, the synthesis and destruction of organic compounds, the absorption and emission of chemical substances by various organisms, as well as many other processes which are in a complex dynamic equilibrium.

The study of the ocean's content of the main biogenous elements such as oxygen, carbon, calcium, nitrogen, phosphorus, and silicon is of great importance.

The Institute's researchers have disclosed the processes of these substances' rotation, evaluated their quantitative transformations, and determined the possibilities for biomass synthesis under various conditions. They have established, for instance, that the low biological productivity of the subtropical and tropical regions of the World Ocean is explained by the shortage of non-organic nitrogen, phosphorus, and silicon compounds in its fifty-meter layer. This finding is of utmost importance for the development of mariculture.

Of no less importance is the study of the migration of certain heavy metals toxic to living organisms (mercury, lead, cadmium, zinc, silver, etc.) as well as the study of oil pollution. The Institute has developed, and patented in many countries, efficient methods of neutralizing oil spills on site. Pesticides of the chlorinated hydrocarbon group (DDT and its derivatives) are also a serious danger. Most of these compounds are insoluble in water, but readily dissolve in organic solvents. This facilitates their wide dissemination in the Earth's biosphere, including the Antarctic fauna.

A direct connection has been established between the level of water pollution with pesticides and the content of pesticides in the fatty tissues of fish. Pesticide concentration in animal tissues increases in the consecutive links of the food chain and may be by one or two orders of magnitude higher than that in water. Thus, further uncontrolled use of pesticides on the continents is extraordinarily dangerous. The Institute's scientists are convinced that the further spread of synthetic surface active substances, with their extensive industrial and domestic application as active detergents, should be drastically restricted. The accumulation of these biologically nondecomposable substances on the surface of the ocean will cause a deterioration of its oxygen productivity.

One of the most fruitful concepts in the hydrochemistry of the ocean is that of its chemical structure and the principles of chemical and oceanographical zoning based upon it. The staff of the Institute has undertaken several projects in the chemical and oceanographical zonation of the oceans, horizontally and vertically, in a depth range between zero and 5,000 meters, divided into ten levels for which hydrochemical maps were compiled to account for oxygen, nitrates, phosphates, and silicon. They proved that the distribution of biogenic elements and oxygen in the ocean is characterized by a fine structure, and that this structure is similar to the fine temperature and salinity profile of the water. They also investigated the variation of the hydrochemical state of the ocean on a variety of scales.

Marine Meteorology

Meteorological observations are usually carried out in accordance with special programs if they investigate the interaction of ocean and atmosphere. Professor V. Samoilenko has done a great deal to provide a methodological foundation for these observations from the research vessels of the Institute. He established a procedure, which has now become a tradition, to include detailed hydro-meteorological descriptions of the regions visited by expeditions in their reports. In recent years, the generalization of a great many hydro-meteorological and oceanological data has led to some novel notions of the climate of the oceans. The close ties between Soviet oceanological studies, and meteorological research in the physics of the atmosphere and climatology, stem from a profound understanding of the nature of the interaction of the hydrosphere and atmosphere. The approach has resulted in a series of important studies that led to a rationale for a further participation of Soviet oceanologists in several international and national programs for the investigation of the part played by the ocean in changes of the global weather and climate. The Shirshov Institute of Oceanology and the Institute of the Physics of the Atmosphere of the U.S.S.R. Academy of Sciences effectively cooperate, since they have much common ground to cover. The drawing up, by Monin, Zilitinkevich, Chalikov, and Kagan of the Leningrad Department of the Institute of Oceanology, of a global numerical model of interaction between ocean and atmosphere has made it possible to begin to assess and correct our notions of the influence exerted by the ocean on the world climate.

Deep-Sea Diving

Before the middle of this century, most studies of the ocean had been carried out from surface ships, with the help of various instruments lowered to the ocean floor or to a certain depth. Later, efforts were undertaken to dive into the depths of the oceans by the French explorer Jacques-Yves Cousteau. The Soviet Union and other countries began building autonomous submersibles somewhat later. Among these craft were submarines (Severyanka class submarines in the U.S.S.R.); depth boats, and underwater habitats, etc.

Starting in the 1970s, the Institute has been engaged in an extensive program of designing, building, and testing various types of underwater craft. At the same time, a team of divers started training for underwater missions. The young scientists A. Sagalevich and A. Podrazhansky undertook the first of these missions.

The Institute has employed submersibles of the *Pisces* and *Argus* series for geological, hydrophysical, and biological research beginning in 1977. Craft of the *Argus* series were developed and built at the specialized design bureau of the Institute. *Pisces* craft built for the Institute in Canada, capable of descending to a depth of 2,000 meters, are largely used to explore the rift zones of the planet. In 1982, the divers of the Institute explored the ocean floor in the vicinity of the Reykjanes underwater ridge. Earlier, they had explored the Red Sea and Lake Baikal and produced some very interesting results. They obtained valuable data about the geological structure of the Red Sea rift and the world's deepest freshwater lake. In 1988, the Institute obtained two new submersibles, built in Finland, with steel pressure hulls, capable of descending to a depth of 6,000 meters—a significant supplement to a world fleet of this kind which includes *Sea Cliff* of the United States, *Nautilus* of France, and *Shinkai* of Japan (all of these with titanium pressure hulls).

Remote-controlled tow-craft are also employed for the exploration of large areas of the sea bottom with the help of television, photo, and bottom-sampling equipment.

Installed at the Southern Department of the Institute and on board the new *Vityaz* (she sailed on her maiden voyage to the Mediterranean in 1982), are pressure chambers where divers are trained and medical and biological studies are carried out. In cooperation with medical institutions, the Institute has reached, in pressure chambers, the level of 45 atmospheres in the heliox (helium-oxygen) breathing mixture (the world record of 68 atmospheres belongs to Duke University in the USA), and the level of 25 atmo-

spheres in the neon-oxygen breathing mixture with density equal to that of heliox at 125 atmospheres. This last experiment indicates, in principle, the physiological possibility for a human being to dive in the ocean without submersibles to a depth of 1,250 meters using heliox or, perhaps, a hydrogen-heliox breathing mixture.

"Field Work": Some Examples

Let us now present a few examples of the Institute's actual "field work" in the ocean.

One of them is the now famous Soviet-American joint expedition POLYMODE (a combination of the Greek-Russian word *Polygon* and the American MODE (Mid Ocean Dynamics Experiment) in the Sargasso Sea on the southern periphery of the Gulf Stream during the period from July 1977 to September 1978. It consisted mainly of a Soviet array of nineteen buoy stations with four current-meters and thermometers at fixed depth on each of the moorings, and the American SOFAR (Sound Fixing And Ranging) system with twenty-five free-drifting neutrally buoyant floats at two fixed depths tracked by using an acoustic signal received by coastal stations.

The main result was the detection of thirteen synoptic vortices in the open ocean passing the area of the experiment with horizontal scales of several tens of kilometers, vertical scales of the order of one kilometer a day and velocities of water motion in the eddies much greater than those of mean currents. The discovery of synoptic eddies in the open ocean was perhaps the greatest event in post-war physical oceanography, radically changing the previous image of a slowly and smoothly flowing water mass of the ocean. The results were published in the joint POLYMODE Atlas (1986) edited by A. Monin (U.S.S.R.) and A. Robinson (U.S.A.).

Unfortunately, joint Soviet-American "field work" in the ocean was suspended during the Carter and Reagan administrations and *Polygon* studies of the oceanic vortices were continued by the Institute alone. One should mention the *Mesopolygon* expedition in the Atlantic trade-wind zone in 1985, with an array of 76 moorings carrying 215 current-meters in an area of 70×80 miles, where a moving and anticyclonically rotating lens of the Mediterranean water was detected at a depth of 800–1,300 meters, with a diameter of about 30 miles, temperature anomaly of $+4°C$, and salinity anomaly of $+1$ promille. The last of this kind of expeditions was the *Megapolygon* expedition in the north-western part of the Pacific to the north of the

A submersible of the U.S.S.R.
Academy of Sciences after a
dive. Since the 1970s, the
Shirshov Institute has been
engaged in an extensive
program of designing,
building, and testing various
types of underwater craft, and
training divers for deep-sea
missions. Photo: Shirshov
Institute

Manned submersible of the *Argus* class. Craft of the *Argus* class were developed and built at the specialized design bureau of the Institute. Photo: Shirshov Institute

Kuroshio current in 1987, with an array of 177 moorings carrying 440 current-meters in an area of about 500×500 kms. The array detected simultaneously a dozen vortices in the upper layer of the ocean, and rather strong synoptically changing currents at the depths of 1,200 and 4,500 meters.

To give an example of marine geological "field work," let us mention the RSSR (Red Sea Submersible Research) expedition of 1980, with the *Pisces* submersible, where a chaos of basaltic pillow lavas was discovered and studied in the axial zone of the bottom of the Red Sea, and dives down to the surface of the hot brine pools (temperature of 60°C, salinity of 30%) were performed, showing the unforgettable picture of "surf" at the "shores" of the pools at a depth of about 2,000 meters.

Our last example will be the biological Amazon expedition of 1983, when the ship *Professor Shtockman* went 2,000 kilometers up the Great River studying the fish and making detailed measurements of the optical properties of "white," "black," and "clean" waters.

Submersible of the *Pisces* class. Built in Canada, *Pisces* are capable of descending to a depth of 2,000 meters, and are used mainly to explore the rift zones of the planet. Photo: Shirshov Institute

Conclusion

No discipline can advance with any success unless it has vehicles for the exchange of information, especially periodicals. The work of the Institute, which the Nauka Publishers have been issuing since 1946, largely paved the way for publishing specialized periodicals concerned with oceanology. All the listed periodicals are published in the United States in English.

In conclusion, one should stress that the staff of the Shirshov Institute pays careful attention to various anthropogenic changes in the ocean. Researchers are making strenuous efforts to solve the practical problems of ensuring safe shipping, expanding ocean fisheries, growing more offshore sea crops, and developing the underwater deposits of oil, gas, nodules, and sulphides of metals.

Oceanologists are facing many pressing tasks. The problems they solve benefit all mankind. What should be done to keep the oceans clean in spite of its intensive industrial development? How can the food resources of the oceans be replenished to a degree corresponding to the rate of their consumption, which is steadily growing? How can climatic changes be forecast to head off any negative impact they may have on farming, industry, and transport?

Soviet marine researchers view the World Ocean as an integral whole and as an extremely complex natural feature, whose natural internal processes and outside influences are all closely interconnected. It takes a special kind of scientific vision to discern in that maze of intertwined relations those which hold the promise of a major breakthrough in several avenues of research at the same time. Many Soviet oceanologists have this kind of vision. They also are sure that the study of the World Ocean in the interests of humanity requires the cooperative efforts of all the countries of the world.

Tow craft. Photo: Shirshov
Institute

7.

The National Institute of Oceanography, Goa

T.S.S. Rao

Horseshoe crabs being studied as an important source for a very sensitive reagent. Photo: NIO

Introduction

India and the Indian Ocean are inseparable. Their histories are intertwined and extend back to the days of the Sumerians, Phoenicians, and perhaps Harappans, who traded among themselves across the Erythrean waters (the present Arabian Sea). From antiquity to the modern laboratories of the National Institute of Oceanography at Goa is unquestionably a quantum jump. This chapter tries to retrace the pathways, both ancient and modern, along which India has traveled to reach today's eminence in oceanographic research in its part of the world.

Vedic Times

Beginning in Vedic times (2500–1500 B.C.), we see in Indian literature many references to the sea, or *sagara*, as the repository of riches of all types, and particularly of *amrita*, the beverage of immortality. There is a fascinating account of groups of beings, the *devas* and the *asuras*, who churn the ocean using a gigantic snake as churning rope and a mountain as churner, to obtain *amrita*. This story, popularly known as *Samudra Manthan*, is the subject of many paintings and sculptures found in the temples of India. As a result of this churning, the story says, the ocean brought forth many gifts to both these groups of beings.

This tale may serve as an indication of the awareness the ancient Indian people had of ocean resources.

Vedantic literature does not contain much specific information concerning the oceans. But there are references in the two epics, *Ramayana* and *Mahabharata*. In *Ramayana*, there is the story of Hanuman, the monkey god, flying from the tip of the Indian subcontinent (near present-day Rameswaram) to Sri Lanka, in search of Sita, the wife of Rama. The author of this epic, Valmiki, gives a beautiful description of the aerial view of the mighty ocean separating the two lands.

In *Mahabharata*, there is the story of the disappearance of Dwarka, the capital city of Krishna, under the sea, described with marvelous precision.

There is clear evidence that the ancient Hindus knew about the monsoons and the rhythm of tides and waves. The discovery, at Lothal (a small town in Gu-jarat), of a dockyard provided with flood gates and ship anchors, bears witness to their knowledge of local tides, which they fully utilized for both navigation and the anchoring of ships (2450–1400 B.C.).

The concept of tides in ancient India is the subject of a paper by Panikkar and Srinivasan: "The Concept of Tides in Ancient India." It would appear that Vedic Aryans knew that the sea swells periodically under the influence of the moon.

"Like the tides of the ocean," the *Maitrayani Upanishad* says, "the approach of one's death is hard to keep back," meaning that the force of the ocean tide, as observed by the ancients, was something that could not be stopped.

Foreign Visitors and Expeditions

The Greek navigators who traded with ports in the Gulf of Kutch and along the mouth of the Indus have left behind even more precise information concerning the tides.

Herodotus (450 B.C.) recounts the regular occurrence of tides in the Arabian Gulf (the present Red Sea). Alexander the Great, who advanced to the mouth of the Indus, appears to have made accurate observations on the ebb and flow of the sea, as recorded by Arrian.

With the discovery of the sea route to India, a large-scale trade in spices, silk, ivory, as well as in gold and silver jewelry, was established between Europe and India, particularly along its west coast. The monsoons, which blow regularly and reverse with the seasons in the Arabian Sea and the Bay of Bengal, proved extremely useful to navigators: the southwest monsoon winds would facilitate sailing eastward from the African and Arabian coasts during April to September, and the northeast monsoon winds did the same in the opposite direction from about January to April. The diaries left by the captains of ships plying the Indian seas contain records of the flows of currents and tides in coastal areas.

As far as the Indian people are concerned, despite their early mythological insights, there was no real interest in oceanographic research until after

World War II. Although some Indian merchantmen ventured forth to trade both in the East and West during the past centuries, by and large they never indicated or cultivated scientific interest in the oceans, their depths, or their fauna and flora. Perhaps the main contributing factor to the apathy of the Indian people toward a scientific approach to the oceans is the concept and practice of *Kalapani*, a term declaring the sea to be "bad waters" and any one who sailed on it liable to social boycott, particularly among higher castes. This tradition is responsible for the long-standing Indian ignorance of the sea. It proved disastrous when India was confronted with the invading European powers, who made full use of their knowledge of the sea both for warfare and for trade.

When the British launched their famous Challenger Expedition in 1872, the Indian people were fully involved—as a subject country, with hardly any scientific activity. The universities that were established in India were primarily designed to prepare clerks to assist their British rulers in governing the large country, using the English language. Scientific research was totally suppressed in Indian universities.

While India was thus slumbering, European and American research ships began to cruise the Indian Ocean to gather information on a variety of subjects, such as currents, plankton, fisheries, and seabed resources.

Before 1947, a number of expeditions to the Indian Ocean were undertaken. Quite a few Englishmen, serving in the various departments of the Government of India, made significant studies on the fauna and flora in Indian waters.

In 1881 a wooden paddle steamer of 580 tons, named *Investigator*, was launched by the Marine Survey Department of the Government of India. This ship received much of the scientific equipment from the *Challenger* for work in Indian waters. The R/V *Investigator* continued its work until 1908, when it was scrapped and a new ship, *Investigator* II, was launched. It was on this ship that Sewell (1925–38) did his important oceanographic work in the Bay of Bengal.

Oceanographic data in the Indian Ocean were collected on a number of expeditions and incorporated in their reports. But the *Sealark* (1905–09) and *Mabahiss* (1933–34) expeditions were the only ones that spent considerable time in detailed investigation of parts of the Indian Ocean. The reports of these expeditions and the papers published by Sewell in the *Journal of the Royal Asiatic Society of Bengal* constitute the only important source of oceanographic data for the Indian seas during that period.

The International Indian Ocean Expedition (IIOE) and Indian Efforts 1947–1966

Accumulation of knowledge of the Indian seas until the dawn of Independence in 1947 was in the hands of India's European rulers. In fact, when Anton Brunn visited some of the Indian universities and showed a movie of his achievements in the Galatea expedition of 1952—of which he was the leader—Indian scientists were mesmerized to hear him speak about the deep-sea fauna in areas quite close to India.

In the same year of Independence, 1947, the Ministry of Food and Agriculture of the new Government of India established a Central Marine Fisheries Research Institute at Mandapan, under the directorship of Dr. H.S. Rao (1947–48), followed by Dr. N.K. Panikkar, who was to play a pivotal role in the development of marine scientific research in India in later years. Meanwhile, the Indian Navy set up a Physical Oceanographic Research Center at Cochin. Anticipating the manpower requirements of the expanding fisheries and oceanographic research in India, many coastal universities, such as those at Waltair, Madras, Annamalai, Trivandrum, and Cochin, opened marine biology departments. But it was not until 1960 that the Federal Government appointed an Indian National Committee on Ocean Research (INCOR) to serve as a focal point for oceanographic research and development in India. Among other things, INCOR was responsible for the drawing up of a co-ordinated plan for India's participation in the International Indian Ocean Expedition. It also was responsible for liaison with international organizations such as UNESCO, the Scientific Committee on Ocean Research (SCOR), and the Intergovernmental Oceanographic Commission (IOC), particularly during the years of the International Indian Ocean Expedition (1960–65).

Those years, of which Daniel Behrman gives a fascinating account in his book *Assault on the Largest Unknown*, were exciting ones in the history of Oceanography.

The first meeting of SCOR took place at the Woods Hole Oceanographic Institution on August 28–30, 1957, under the presidency of Roger Revelle, who was then the Director of the Scripps Institute of Oceanography. At that meeting it was decided to launch an International Expedition into the Indian Ocean. Robert Snider was appointed Coordinator for the Expedition. He did a remarkable job in enlisting the cooperation of various countries and was able to merge all efforts to make the IIOE a great success. In India he was able to gain the attention and approval of Jawaharlal Nehru, then Prime Minister, through his right-hand man Dr. Homi Bhaba. In the United States, he succeeded in having IIOE endorsed by President Eisenhower and, subsequently, by President Kennedy.

The list of scientists involved in the planning and execution of the Expedition (of which this writer was privileged to be part from the very beginning), is quite long and includes the most distinguished names in oceanography of this century. Suffice it to mention G.E.R. Deacon, Director of the U.K. Oceanographic Institute, Columbus Iselin, Director of the Woods Hole Oceanographic Institution, Gunther Boehneke, head of the German Hydrographical Office (who later served as first Secretary of the SCOR), Lev Zenkevich, Soviet marine biologist, and George Humphrey from Australia. All were united in the hope that the IIOE would fill some of the large gaps in our knowledge concerning the Indian Ocean and encourage the development of oceanographic centers in those countries bordering it.

Some forty ships took part in the Expedition. They came from thirteen countries: Australia, France, the Federal Republic of Germany, India, Indonesia, Japan, Pakistan, Portugal, the Republic of South Africa, Thailand, the United Kingdom, the United States, and the U.S.S.R. It was agreed that zooplankton samples on all ships would be taken by an Indian Ocean Standard Net and sent to the Indian Ocean Biological Center at Cochin. Similarly, meteorological data were pooled for processing at the Meteorological Center at Colaba, Bombay. These two Centers were established by the Indian Government to serve the international scientific community and facilitate coordination and integration of the Expedition's work.

This turned out to be extremely fortunate for India, and greatly stimulated exchanges of ideas and interest in—even enthusiasm for—oceanographic sciences.

The International Indian Ocean Expedition brought India closer to the Indian Ocean than anything else that had happened during its long history. Even when she was conquered and subjugated by Dutch, Portuguese, French, and English coming by the sea route, India did not awaken to the need for knowledge concerning the Indian seas and its mastery. It was only when many countries, from far away in Europe and America, sent their research ships such as the American *Anton Brunn* and *Atlantis*, the British *Discovery*, the German *Meteor*, the Russian *Vityaz*, and many others, that Indians both at the political and scientific levels,

Samudra Manthan—or "the churning of the oceans." Ancient temple painting. Courtesy: T.S.S. Rao

though perplexed at first, began to realize the importance of ocean studies.

From 1960 to 1962, the *Vityaz* made the first systematic study of the geology of the seabed of the Indian Ocean. Together with the observations made by HMS *Owen* and *Darlymple* in 1961 and 1963, followed by three cruises of R/V *Argo* and R/V *Horizon* from Scipps and four cruises of R/V *Vema* from Lamont (1959–64), details of the Indian Ocean seabed became clearer—particularly the delineation of the mid-Indian-Ocean Ridge and its northward extension into the Arabian Sea as Carlsberg Ridge. A new, 90° East Ridge was also discovered. This ridge has been described as perhaps the straightest feature on the face of the earth, emerging from the sediments of the Bay of Bengal, a thousand kilometers north of the equator, and running south in a line 4,800 kilometers long near the meridian for which it was named by Heazen and Tharp.

The Gulf of Aden was identified as an embryonic ocean connected to a ridge system in a continuation of the great Rift Valley of Africa. The Gulf is assumed to be moving away from Africa and Arabia at the rate of 2 centimeters a year.

Perhaps the most fascinating and epochal finding was made by *Discovery* and *Atlantis* in the Red Sea. *Atlantis* II also joined these studies.

These studies indicated that the Atlantis II Deep (a name given to a "graben" in the Red Sea investigated by *Atlantis* II earlier) had amazingly high temperatures at about a 2,500 meter depth: as high as 55.9°C, while the average temperature was no more than about 30°C. The water was acidic and anaerobic, i.e., with no trace of oxygen. On the other hand, it contained amazing quantities of heavy metals. When the sediments in these deeps were studied, they were seen to exhibit a multitude of colors: white, black, red, green, blue, and yellow, mainly containing iron, manganese, zinc, copper, and traces of silver and gold. Eight of the samples cored by *Chain* were analyzed by F.T. Manheim of the United States Geological Survey and J.L. Bischoff at Woods Hole for their economic value. This was estimated to total 2.3 billion 1968 dollars, including $780 million in zinc, $1,100 million in copper, $280 million in silver, and $50 million in gold. Until the collapse of mineral prices in the eighties, the Saudi-Sudanese Red Sea Commission was planning to exploit these mineral resources with West German technology. Like other seabed mining projects, this one is now "on the back burner," but, undoubtedly, its day will come.

In the field of marine biology and fisheries science, the results of the IIOE were highly encouraging. Based on productivity data, there was a prediction that Arabian Sea fisheries alone would be of the order of ten to twenty million tons per year, as against the present harvest of about two million tons.

Perhaps the most important fallout from the results of the IIOE was the establishment of a link between the atmosphere and the ocean in the generation and distribution of the monsoons over India. At one time, five research aircraft were flying out of Bombay under the direction of Professor Colin Ramage. They collected very valuable information about monsoon conditions both in the air and immediately above the sea surface. The results were published in two volumes of atlases dealing with the meteorological aspects of the expedition. The first volume contains 144 charts showing the surface climate of 1963–64 and is based on 194,000 ship observations. These charts give, for each 5° square, data on wind, pressure, air, and sea surface temperature, vapor pressure, clouds, precipitation, and heat exchange. The second volume gives data on the upper air climate of the Indian Ocean and its adjoining continents, and is based on 750,000 balloon ascents made at 274 stations operated by 45 countries, 118,000 wind measurements by crews of 32 airlines, airforce planes, and research ships. With all these observations we now have a much better understanding of the monsoons, and these data form a very good base to build prediction models of the monsoon circulation over the Indian landmass.

Three more sets of atlases were compiled as a result of the IIOE: on the hydrography of the Indian Ocean, on the zooplankton distribution, and on the primary production in the Indian Ocean. No other ocean can claim such a broad coverage.

It is clear that the twenty-year period, from Independence in 1947, to the formal inauguration of the National Institute of Oceanography in Delhi in 1966, was of crucial importance for the development of oceanographic science in India and the Indian Ocean area.

Not only on the international plane, but also within the national context, there were a number of important developments.

The Central Marine Fisheries Institute at Mandapan, devoted to marine fisheries research and the survey of exploitable fishery resources, soon mushroomed into a very large organization employing about 500 scientists and technologists in regional units located all along the east and west coasts of the Indian subcontinent. The Norwegian government provided aid in the form of a research vessel, named *Varuna*, of about twenty meter length, which did excellent survey work in the Arabian Sea and the Bay of Bengal. Fish landings, which amounted to 50,000 tons in 1947, increased to 800,000 tons by 1960. This was mainly due to improved mechanization of the fishing trawlers and an extension of the fishing area, based on fisheries surveys.

In addition to its work on fisheries, the Institute had other groups working on plankton and productivity, on physical and chemical oceanography, and on statistics. It also published two journals: the *Journal of the Marine Biological Association of India* and the *Journal of Fisheries*.

Soon after the establishment of the Central Marine Fisheries Institute, a number of other institutes were founded in the fifties and sixties: the Central Institute of Fisheries Technology and the Central Institute for Fisheries Operatives and Nautical Training, both in Cochin; the Central Institute of Fisheries Education, at Bombay; and, also at Bombay, an offshore fisheries survey unit, established by the Ministry of Food and Agriculture, exclusively for exploratory fishing.

While Government attention was mainly directed toward fisheries research and development—and this quite rightly so, considering the expanding food requirements of the growing Indian population—the coastal universities concentrated their efforts on creating a pool of marine biologists to meet the recruitment needs of the fisheries organizations. In this process they neglected other areas of oceanographic research, such as physical, chemical, and geological oceanography—and research in marine geophysics and marine meteorology were almost unknown. Andrah University, among the coastal universities, stands out for the diversification of its studies into all these branches as early as 1952. This development was greatly facilitated by the arrival on the scene of a Fulbright Professor, E.C. La Fond and his wife Catherine, from the United States.

During their two visits to Andrah University, in 1952–53 and 1956–57, a total of about fifty oceanographic cruises were conducted in the Bay of Bengal, using Indian naval ships, and a large number of papers were published.

At the same time, the laboratories of Annamalai University, located at Porto Novo, began their studies of mangroves and estuarine oceanography. In recognition of their contribution, these laboratories were upgraded to the status of Centers for Advanced Studies in Marine Biology and provided with extra funds.

The zoological laboratories at Madras had pioneered research in marine biology in the early thirties,

and N.K. Panikkar had done important work on the osmoregulation in crustaceans. In later years these contributions to research dwindled somewhat. However, the Madras museum has one of the best marine collections in the country. It also has an excellent library, the Connamera Library, named after a British Governor of the Madras presidency.

Both Trivandrum, which has no harbor, and Cochin, with an excellent harbor and extensive backwaters, became meccas for marine biologists and fishery scientists from all over India. Cochin, in particular, has perhaps the largest concentration of manpower in the world dealing with fishery science and oceanography.

The backwaters of Cochin, which are extensive and highly productive, became the subject of important studies by a large number of scientists led by S.Z. Qasim, who was appointed in 1962 as scientist in charge of the International Biological Program administered by the Indian Ocean Expedition Directorate. The cumulative effect of these systematic surveys of the Cochin waters resulted in the promotion of the export of shrimp and prawns which today earns 300 million U.S. dollars or more for the country. Qasim later became Director of the Central Marine Fisheries Research Institute and, after Panikkar's retirement, he took over as Director of the National Institute of Oceanography. In both these positions he achieved remarkable progress in practically every field of oceanography. Apart from being a meticulous scientist, his engaging personality attracted the attention of the media, sometimes very much to his chagrin. Qasim, perhaps, was the first to produce a factual image of the seasonal changes in various environmental factors in relation to gross and net primary production in a tropical estuary in India. He spent the final years of his active life as Secretary of the newly established Government Department for Ocean Development (1981–89).

Oceanography thus was well advanced, and the stage was set for the Government's decision to establish a National Institute of Oceanography.

The Birth of The National Institute of Oceanography

In 1963, a feasibility study for the establishment of a National Institute of Oceanography was presented to the Indian National Commission for Oceanographic Research. The study was unanimously approved and the Institute was to be founded by pooling the already

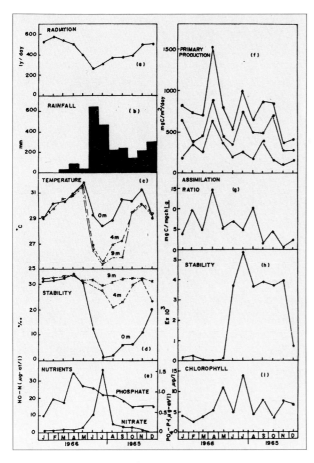

Organic production in a
tropical estuary. Courtesy: NIO

existing scientific and administrative staff of the Indian Ocean Expedition Directorate and the Indian Ocean Biological Center—altogether about fifty scientists and supporting staff, who formed the nucleus of the new Institute that was formally brought into being in 1966 by the order of the Central Government, as one of the national laboratories. It was the last laboratory to be inaugurated by that great thinker and humanitarian, Pandit Jawaharlal Nehru, Prime Minister of India.

The Aims and Objectives of the Institute

The Institute's mandate is comprehensive. It is to develop knowledge related to the physical, chemical, biological, geological, geophysical, and engineering aspects of the seas around India. And this includes the exploitation of the living resources of the sea; sea farming technology; deep-sea exploration for minerals; drugs from marine plants and animals; utilization of energy from the sea, such as wave energy or Ocean Thermal Energy Conversion; development of offshore oilfields; coastal zone and harbor development; and studies for effective pollution control. In addition, the Institute is to develop self-sufficiency in marine instrumentation, as well as data and information handling.

The Institute began its operations in New Delhi, but was shifted to Goa in 1969 and moved into its present buildings in 1974. During this period of about nine years, the Institute expanded its activities and recruited more staff. Today it has nearly 600 people on its rolls—about 300 scientists, with 300 technicians and administrative staff. It operates two research ships.

The Institute is located on the Dona Paula Plateau, seven kilometers from Panjim City and overlooking the Arabian Sea—an appropriate setting for an Oceanographic Institute. The buildings, occupying a total area of 121,387 square meters, were planned for completion in three phases. The first phase of the building, with a plinth area of 6,300 square meters, was completed by 1973, and that was when the Institute was transferred there from a hitherto rented building at Miramar. The second phase of the building was completed in 1985 (1,700 square meters). During the present, third, phase an auditòrium, museum, aquarium, conference hall, library, and administration building are being added.

The research and development work is carried

A view of the National
Institute of Oceanography at
Dona Paula, Goa. Photo: NIO

out by nine divisions: Physical, Chemical, Geological, and Biological Oceanography; Instrumentation; Ocean Engineering; Biofouling and Corrosion; Information and Planning; and Data and Computer.

All the divisions participate in applied research, sponsored mainly by industries and government agencies, and thus basic research is accorded a low priority; there is increasing support for universities to undertake such studies in collaboration with the NIO. The research ships are utilized for major work programs sponsored by the Navy and the Department of Ocean Development for their manganese nodule survey programs, etc.

All the laboratories are well equipped with the most up-to-date analytical instrumentation. There is also a very good workshop, a modern library with reprographic facilities, a photographic room, drawing room, etc.

Research Vessels: the R/V *Gaveshani* and the O/R/V *Sagar Kanya*

The R/V *Gaveshani* (a Sanskrit word meaning "in search of knowledge") is the first oceanographic ship acquired by the NIO. The ship was built in Scotland in 1965 as hopper barge for doing dredging work in the Port of Calcutta. It was purchased by the NIO for about U.S. $600,000 and converted at an additional cost of about $1.5 million to a fully operational oceanographic ship. It has passive tank stabilizers and heavy duty winches for geological work. It has four multipurpose laboratories, into which equipment is fitted for each cruise.

So far, this ship has served NIO extremely well and has completed 165 cruises covering more than 200,000 kilometers. This ship was instrumental in locating and recovering manganese nodules in the Indian Ocean in 1981 (see below).

The *Sagar Kanya* is India's second ocean-going research vessel, and perhaps one of the most modern research vessels in the world today. The name, too, is Sanskrit and means "daughter of the sea." This ship was built at Lübeck-Travemünde in the Federal Republic of Germany under the Indo-German Economic Development Cooperation Agreement and cost about U.S. $40 million. The vessel was delivered to the Department of Ocean Development of the Indian Government on March 25, 1983, at Travemünde, by the shipbuilders M/S Schlichtingwerft, and was received by S.Z. Qasim, Secretary of that Department. It left immediately for India on a trial cruise under the leadership of this writer and reached Goa on June 27, 1983,

when it was handed over to the Institute for management.

The vessel has a total of thirteen laboratories spread over three decks, and various other additional work spaces. The laboratories are fully equipped with the basic instrumentation for oceanographic research in geology, geophysics, meteorology, chemical oceanography, physical oceanography, and marine biology. In fact, this ship is a floating oceanographic institution and meets the requirements of all types of work in the sea, and the turn-over period between cruises is minimal.

Major Contributions of the NIO

During the last twenty-two years, the Institute has greatly developed and expanded. It is now perhaps the only one of its kind on the Indian Ocean seaboard.

. . . Pipe-line Surveys

The pipe-line survey for the Oil & Natural Gas Commission (ONGC) was the first major work undertaken by the Institute during the seventies. Working on the R/V *Gaveshani*, the scientists of the Institute produced a seabed map of great precision enabling the ONGC to lay their oil pipe from the Bombay High offshore oil rigs to Bombay, over a distance of nearly 160 kilometers. This work was commended both by the ONGC and their foreign consultants, and the standing of the NIO was well recognized. As a result, more and more industries with marine survey problems began to approach the NIO for assistance. These include effluence disposal points, harbor construction and reclamation, living resources survey, pollution monitoring, problems of erosion, aquaculture, etc.

. . . Seasonal & Inter-Annual Variability in the Northern Indian Ocean

Oceans are the flywheel of the world climate system. A monsoon draws its energy from the ocean. The heat content in the upper layers of the ocean plays a predominant role in the exchange of energy between oceans and atmosphere. Physical oceanographers of the Institute are concerned with studies of the variability of heat budget parameters in the northern Indian Ocean. They are trying to delineate various processes such as advection (local change of atmo-

The research vessel *Gaveshani* has completed 165 cruises, covering more than 200,000 kilometers. Photo: NIO

The research vessel *Sagar Kanya*. Built in Germany, the *Sagar Kanya* is one of the most advanced research vessels in the world—a floating oceanographic institution.
Photo: NIO

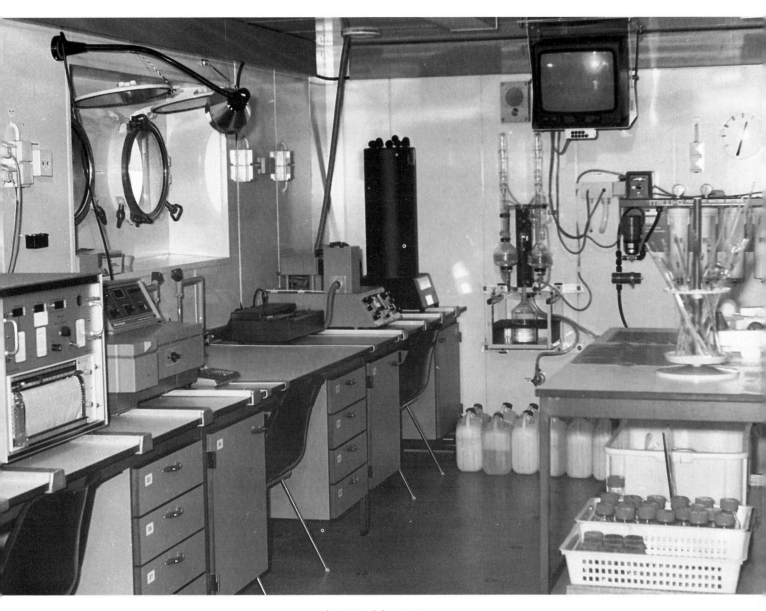

Chemistry laboratories on
board the *Sagar Kanya*. Photo:
NIO

spheric temperature), evaporation, net radiation, mixing, diffusion, etc., in controlling the heat content of the upper ocean.

Recent studies have shown that the sea surface temperature increases from about 25°C in the northern Bay to about 29°C in the southern region during the northeast monsoon season. In contrast, the temperature gradient reverses during the southwest monsoon season, when high sea surface temperatures, exceeding 30°C are noticed at the head of the Bay and cold water below 27.5°C in the south. The cyclone heat potential is conspicuously high in the eastern Bay of Bengal surrounding the Andaman-Nicobar Islands during the southwest monsoon season. During the post-monsoon season, the cyclone heat potential is high in the southern Bay of Bengal.

These studies, yielding a mass of data, are fundamental to an understanding of the genesis—and, therefore, the prediction—of cyclones.

... Remote-Sensing of Oceanic Parameters

Oceanic parameters such as sea surface temperature, chlorophyll concentration, humidity profiles, etc., can be measured by satellites. The practical application of remote-sensing lies in predicting the fisheries potential, monsoons, forecasting waves, climate, mapping coastal vegetation for its conservation and management, monitoring pollution, and evaluation of mineral resources. Initial studies have been conducted along the southwest coast of India, and remotely-sensed data compared well with data collected by traditional methods. An image processing system has been acquired by the Institute for processing satellite images and related data.

... Bioactive Substances from the Seas Around India

Many marine animals and plants exhibit bioactive properties that can be utilized as a source for organic chemicals and drugs. Extracts of more than 172 plant and animal species have been prepared and screened. More than 52 species have shown promising pharmacological activities, including antifertility, diuretic, antidepressant, antispasmodic, or analgesic action. The structure of the isolated active chemicals is determined by using sophisticated instruments such as nuclear magnetic resonance (NMR), infrared (IR), mass spectrometer (MS), etc. Drugs from the sea is a field of major potential for the future, and offers new avenues

Balloon-launching deck on the research vessel *Sagar Kanya*.
Photo: T.S.S. Rao

of economic growth to developing countries bordering tropical seas.

...Marine Pollution Monitoring

Increasing industrialization and waste disposal in the estuaries and along the coast have necessitated monitoring and control of pollution to avoid harmful effects on the marine flora and fauna and, directly or indirectly, on human populations as well. Pollutants have been identified for their composition, distribution, and concentration. Oil is a major pollutant, but its effects are restricted to a few major harbors. The NIO has completed a number of studies on industrial, agricultural, and domestic-waste pollution and formulated recommendations to the government and various industries for the discharge of their effluents to minimize marine pollution.

...Sedimentological Studies

Different types and aspects of the sediments on the continental margins of India have been studied and significant results have been obtained. These include the exploration of the placer deposits along the Konkan coast (Maharastra), which contain ilmenite, an important source of titanium. Geochemical maps showing the distribution of various elements have been prepared. These are useful in interpreting the environmental conditions for mineral formation, planning for future mineral exploration and offshore construction.

Paleoclimatic studies have shown that the climate of peninsular India about 10,000 years ago was semi-arid and that the change to the present moist conditions was geologically sudden and associated with a low sea level, about twenty meters below the present sea level.

...Polymetallic Nodules—India's Work on the Resources for the Future

The floor of the Pacific, Indian, and Atlantic Oceans is spread with polymetallic nodules which, in some zones, have a high concentration of valuable metals such as copper, nickel, cobalt, and manganese. The highest concentration of these nodules has been found at depths ranging from 3,500 to 5,000 meters and covering a total area of 46 million square kilometers in the World Ocean. Of this area, some 23 million square kilometers are in the Pacific, 15 million in the Indian Ocean, and 8 million in the Atlantic Ocean. The total

Stern view of the *Sagar Kanya*. This ship has thirteen laboratories, spread over three decks, and various other work areas. Photo: T.S.S. Rao

mass of these nodules is estimated to be over 1.7 trillion tons, of which the Pacific Ocean has 1.5 trillion, the Indian Ocean 0.15 trillion, and the Atlantic some 0.05 trillion tons.

Polymetallic nodules vary in color from light brown to black, and they come in different shapes. Some are spherical, others are discoidal or flat. They vary in size, usually ranging from 1 centimeter to 25 centimeters in diameter. The average size of a nodule is about 5 centimeters in diameter. Occasionally nodules are found as agglomerates. The diameter of such agglomerates sometimes exceeds one meter; some of them weigh several kilograms. The nodules are porous and relatively light. Their cross-section shows concentric rings. Usually they have one or two nuclei around which the minerals accrete in successive layers. The nuclei often are of volcanic material such as palagonite, or pumice; nodules therefore are often found in volcanic regions. The concentration of nodules at the surface of the sediments has been attributed to the activity of burrowing organisms, or to upward diffusion of manganese, or to seismic activity, or combinations of all three. Nodules are formed on the ocean floor, usually at a rate of a few atomic layers per day. In a way they are a renewable resource, for they keep forming at the rate of 10 to 20 million tons every year. The resources of nickel, manganese, cobalt, and copper they contain are estimated to be far more substantial than that which is available on land.

The Indian exploration of polymetallic nodules in the Indian Ocean has been widely reported. Indian interest in the nodules goes back to 1977 when the NIO started a project in this field. In January 1981, the research vessel *Gaveshani* collected samples from the Arabian Sea. India has now identified two mining sites in the Indian Ocean. It also has earned the status of "Pioneer Investor" in recognition of its work. This status was conferred upon India by the Third United Nations Conference on the Law of the Sea in April 1982. India shares this distinction with France, Japan, and the U.S.S.R., and with four multinational consortia. These latter, however, have not yet taken advantage of their status, while the four above mentioned States officially registered their claims in 1987 and obtained internationally recognized exclusive tenure to their mine sites. India was the first developing country to have an impressive polymetallic nodules exploration program, and to be recognized as a "Pioneer Investor." It has now been followed by China, which was awarded "Pioneer Investor" status in 1990, and a consortium of Eastern European States and Cuba—registered as "Pioneer Investor" in August 1991.

. . . Living Resources

The mapping of fertility and productivity potentials of different areas in the seas around India has helped in locating new rich fishing grounds. This was based upon research on primary, secondary, and benthic production.

Aquaculture or sea farming is one of the research areas. The Institute has developed a new technique for rope culture of green mussels on floating rafts that has resulted in extremely high production rates of about 480 tons per hectare per year, with as many as three harvests in a year, and a high rate of return of about 181 percent on a low capital investment. Mass culture of the brine shrimp *Artemia*, which serves as food for a variety of cultured species, has been perfected. The traditional technique of alternating rice and prawn harvests in paddy culture has been improved.

Seaweed and mangroves are useful for food, fodder, fertilizer, and pharmaceutical industries. Intensive surveys of seaweed resources along the coasts of Goa, Maharasthra, and Gujarat have been carried out. Attempts are being made to cultivate economically important seaweed and mangroves. Fertilizer from seaweed has been prepared and has shown good results on many commercially important plant species.

. . . Biofouling and Corrosion

Biofouling is a phenomenon in which plants and animals cause deterioration of materials used in the marine environment, whereas marine corrosion is a process in which various metals and their alloys are degraded by seawater converting them into their pristine form (ore).

Since these problems are site-specific, data are being collected on various aspects of the biofouling and corrosion of different materials at various places on the Indian coast, and from the surrounding waters. The data collected should help to suggest measures for the protection of materials in the marine environment.

. . . Marine Instrumentation

Research and Development in marine instrumentation is emphasized as a means to enhance self-sufficiency and reduce dependence on imported instruments. The Institute has already designed and produced a number of instruments for marine surveys, such as a tide and wave recorder and counter, a depth recorder, an elec-

Mussel-culture experiments. The Institute has developed a new technique for the rope culture of green mussels on floating rafts. Photo: NIO

tronic sedimentation balance, an inductive salinometer, an electromagnetic current meter, an electronic bathythermograph, a telemetering system, a wave and tide gauge, three channel air thermometers, and a CTD (Conductivity, Temperature, and Depth Recorder) system. One of the more important contributions has been the designing, fabricating, and installing of the unmanned weather station at Antarctica, which has given excellent results. The NIO has also developed a system for data transmission via satellite from the research vessels to the data center of the Institute.

. . . NIO's Role in Antarctic Research

In 1981 the first Antarctic Expedition was organized under the leadership of S. Z. Qasim, and several scientists from the NIO and other organizations participated. This was the first time that Indian scientists and technicians were exposed to the harsh climate of Antarctica, and, before setting out on this expedition, they were trained in the snow-bound Himalayas for acclimatization to the cold. The first land station established in Antarctica was named Dakshin Gangotri, as distinct from Gangotri in the Himalayas, where the river Ganges originates.

Antarctica holds a magic attraction for mankind. Hitherto it was the preserve of developed nations, such as the U.S.A., U.S.S.R., Norway, U.K., Japan, and others who had long established scientific stations. Some of them claimed sovereignty over large areas of the Antarctic landmass. The Falkland war between the U.K. and Argentina focused attention on the importance these countries attach to the potential resources in the Antarctic sea and the adjacent landmass.

Meanwhile, there had been another development at the United Nations: the U.N. Conference on the Law of the Sea proposed the powerful principle of the Common Heritage of Mankind to explore and manage the seabed resources of the oceans through a special institution of the U.N. system, called the International Seabed Authority. Scientists and politicians have been studying the possibility of establishing a similar mechanism for the Antarctic continent and its adjacent ocean: a subject that is quite sensitive today in international discussion fora. Considering the fragility of the Antarctic environment, a consensus now seems to be emerging for declaring the Antarctic continent and the surrounding sea to be an International Nature Park—to preserve its pristine character and the integrity of its ecology.

India decided to develop expertise in this difficult area, and has now created a pool of scientists and technical staff for undertaking exploratory work in that far-off continent. India is currently operating the Seventh Antarctic Expedition and has established two permanent winter stations on the Antarctic landmass. In these efforts NIO's role has been pivotal and pioneering.

. . . International Programs

The Institute has been quite responsive to international initiatives for the advancement of oceanographic research in the Indian Ocean. Apart from participating in various international forums such as UNESCO, IOC, UNEP, or UNDP, the Institute has fostered collaboration with many countries, e.g., the U.S.A., Germany, the U.K., Norway, Sri Lanka, Mauritius, Seychelles, Kenya, the U.S.S.R., and others. NIO scientists have successfully participated in a number of international projects such as the Indo-Soviet Monsoon Experiments during the seventies, or the international calibration exercises organized by IOC/ICES in the early eighties. The NIO data center is recognized as a repository for data from the Indian Ocean region by the IOC/UNESCO data group.

An international training program on marine resources management and conservation in the Indian Ocean Basin and Adjacent Seas was organized by the NIO in collaboration with the International Ocean Institute, Malta, from October 4 to December 10, 1982. Elisabeth Mann Borgese, chairman of the organizing committee, planned the program for the developing countries in this part of the world. About thirty renowned experts from the U.S.A., Canada, the U.K., and Europe, and thirty scientists from India delivered lectures followed by seminars and practical exercises. This writer and G.L. Kesteven of Australia were the directors of the program. Trainees came from India, Sri Lanka, Thailand, Malaysia, Vietnam, the Philippines, Kenya, Iraq, and Iran. This program was generously supported by the Department of Ocean Development and the Ministry of External Affairs of the Government of India as well as by UNEP, FAO, CIDA, and UNESCO.

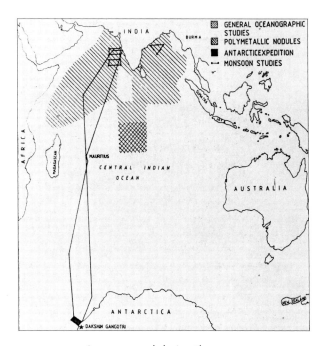

Areas covered during the cruises of the Institute's research vessels in the Indian Ocean. Courtesy: NIO

The training program was most welcome in India, since it created a new awareness among the participating countries with regard to the new Law of the Sea and its impact on marine scientific research, as well as the importance of managing the vast resources of the Indian Ocean.

Other, briefer, workshops on marine sciences were organized in collaboration with Germany, Norway, the U.S.A., the U.S.S.R., and various United Nations organizations, during the following years.

The International Ocean Institute has organized subsequent training programs in cooperation with the Indian Institute of Technology in Madras.

The Future

In recognition of the great and rapidly growing importance of marine resources for India's future, the Government of India created a new Department of Ocean Development and appointed S.Z. Qasim as its Secretary in 1981. A policy paper on Ocean Development was tabled in the Indian Parliament, and this forms the basis for future activities in ocean research and development. A development profile of these activities was also prepared for the period 1985–90 by Qasim and the present writer and submitted to the Planning Commission.

On the basis of these documents, the Department of Ocean Development began to organize its activities in a systematic manner and gave massive support for oceanographic research, particularly at the NIO, for polymetallic nodule surveys and for the Antarctic expeditions. The stage is now set for developing mining and metallurgical systems for the production of the metals from the nodules. This by itself is an enormous task, involving capital and recurring expenditures on the order of at least $100 million over the next few years, besides technical and engineering design and development of the equipment. The lead time for establishing a successful commercial venture for the extraction of metals from the polymetallic nodules is about fifteen to twenty years.

There will be further support for Antarctic research, particularly the survey of the krill resources that appear to be attractive as a rich source of protein for India's growing population. The reports of the International BIOMASS program have raised hopes for the harvesting of krill resources, and India should do well to survey and identify areas of krill distribution and abundance for future use.

Technologies for ocean and seabed surveying have advanced to the point where it is now literally possible to pick up a pin from the bottom of the deep ocean. This was demonstrated most convincingly when scientists from France and the United States retrieved the debris of the Air India jumbo jet that had crashed off the coast of Ireland two years ago, or when parts of the *Titanic*, which had sunk in the Atlantic, were salvaged by using remote-controlled submarines backed by underwater TV scanners. With acoustic scanners (multibeam), it is now possible to map the seabed rapidly and with great precision. It would appear that the whole ocean, covering nearly seventy-one percent of the planet, may now be searched and surveyed.

Against this background, the National Institute of Oceanography at Goa has a tremendous opportunity to diversify and make full use of the Indian Ocean for the benefit of the Indian people and the other countries bordering the Indian Ocean. There is a great need for supporting development of marine instrumentation, ship and buoy design technologies, and modeling techniques, if India and other developing countries are to make progress in resource management of the vast oceans. Each of these countries has an extensive Exclusive Economic Zone—India has about two million square kilometers—whose resources are there to be explored and utilized. Since almost all the countries bordering the Indian Ocean are developing countries, there is an urgent need for organizing a Regional Marine Research and Development facility as a focal point for the exchange of information and strategies in this part of the world.

8.

The State Oceanic Administration, Beijing

Qin Yunshan

The Institute of Oceanology,
Quingdao, established by the
Chinese Academy of Sciences
in 1959. Photo: SOA

Introduction

The People's Republic of China is a marine, as well as a continental country, with a vast territory, a beautiful landscape, and an immense sea area rich in resources. Surrounded by the sea on the east and south, China borders on the Huanhai Sea, the East China Sea, and the South China Sea, in addition to her inland Bohai Sea, covering a total area of over 4.7 million square kilometers. The mainland shoreline stretches north-south over more than 18,000 kilometers. Islands numbering over 6,000 spread all along the coast like strings of pearls inlaid in the sea. The total area of China's coastal zone is about 500,000 square kilometers. Rich in a variety of resources, China's coastal zone offers favorable physiographic conditions, and beautiful vistas.

Since ancient times, China has been utilizing the seas for fishery, salt extraction, and shipping—her prosperity has been linked with these activities. Excavation of ancient relics has proved that as far back as 18,000 years ago, the "Upper Cave Man" living in Zhoukoudian used shells of the sea blood-clams to make ornaments. At the end of the primitive age, 4,000 years ago, residents settling in coastal areas collected a large number of shellfish for food. Wang Chong (Lun Heng) suggested the causes of tides as early as the Han dynasty (22 B.C.–A.D. 221). By the Sung dynasty (A.D. 960–1279), China's ships could carry over a thousand people. The use of the compass, which could give twenty-four bearings, and detailed sailing charts showing the harbors, mileage, and submerged rocks along the route, facilitated early navigation. During the Yuan Dynasty (A.D. 1280–1368) utilization of monsoons and of the Kuroshio Current were taken into consideration in opening a sea route for the shipping of grains from the Changjian river-mouth northward. And, in the Ming Dynasty (A.D. 1368–1644), Zheng He, the famous navigator, led an ocean-going fleet—which was then the largest in the world—and voyaged to and from the Pacific and Indian Oceans seven times. The fleet left its mark in almost every corner of Southeast Asia, traveling to India, the Persian Gulf, the Red Sea, and East Africa—covering a total of over fifty thousand kilometers and visiting more than thirty countries. On their last voyage, they crossed the Indian Ocean and the Red Sea and, moving along the East African coast, discovered the island of Madagascar, not far from the Cape of Good Hope. Carrying more than 27,800 crew and a total load of 150,000 kilograms, the fleet consisted of 62 large ships, each about 147 meters in length and 60 meters in width—a remarkable undertaking in the history of navigation.

China has had a distinguished history of navigation, but the closed-door policy pursued in subsequent eras, content with its focus upon the so-called "vast territory and abundant resources," seriously hampered the development of China's marine sector in more modern times.

After the founding of the new China, the traditional marine industry has recovered and been developed. Marine scientific and technological institutions were rebuilt or newly constructed, and long-term programs for marine investigations and scientific research were formulated. As a result of organizing large-scale cooperation, conducting marine surveys, and developing marine affairs, significant progress has been made in marine investigations and scientific research.

In rediscovering the importance of the oceans, especially through the first national comprehensive marine investigations that took place during the 1958–60 period, and the first coastal zone survey in 1960 (both organized by the Marine Specialized Group of the State Commission for Science and Technology), many problems that urgently needed to be addressed were identified: the safety of ships could not be guaranteed while operating at sea; marine fisheries resources had not been fully and rationally tapped; the stocks and distribution of undersea mineral resources had been poorly understood; and there had been a lack of oceanographic data for purposes of national defense and at-sea operations. There was much to be done, and the necessary infrastructure had to be developed. In 1963, twenty-nine oceanographers proposed that the State Oceanic Administration be established.

On July 22, 1964, the State Oceanic Administration was approved by the Standing Committee of the National People's Congress. Directly affiliated to the State Council, the State Oceanic Administration is now the department of the central government in charge of the administration of marine affairs. Its responsibilities are mainly in the management of law-

Abalone. They are now widely
cultured and popular as a
food source. Photo: SOA

Sea cucumbers, class
Holothuroidae. This class
comprises about 500 species,
ranging from those that thrive
in shallow water to those that
live in the deepest parts of
the ocean. Photo: SOA

enforcement, ocean service, research, and coordination. The work accomplished by the Administration in the two decades since its establishment has been impressive.

Institutions and Staff

The State Oceanic Administration (SOA) is headquartered in Beijing. Under its authority it has established three Branches, seven Research Institutes, a Marine Environmental Forecasting Center, a China Ocean Press, and a special secondary school. In addition, there are sixty ocean stations in coastal areas.

The North Sea Branch, the East Sea Branch, and the South Sea Branch are in Quingdao, Shanghai, and Guangzhou respectively. Each Branch has its own research fleet, coastal station network, and monitoring and surveillance system. Each is responsible for the tasks of marine environmental monitoring, surveillance, and forecasting services, as well as environmental protection and law-enforcement in its respective sea area.

Situated in Quingdao, Hangzhou, and Xiamen are the First Institute of Oceanography, the Second Institute of Oceanography, and the Third Institute of Oceanography respectively.

The Quingdao Marine Biological Laboratory, the forerunner of the Institute of Oceanology, was established as a marine biological laboratory of the Institute of Hydrobiology, Wuhan, in 1950, with the distinguished biologist, the late Professor Tong Dizhou (T.C. Tung) as its founder and first director. The laboratory became an independent research unit of the Chinese Academy of Sciences in 1954, to promote the study of marine biology. It was further expanded into an Institute in 1957.

To keep pace with ever increasing research demands, in 1959 the Chinese Academy of Sciences decided to enlarge the Institute once again, and transformed it into a multidisciplinary institution, that is, the present Institute of Oceanology.

The Institute now has a scientific and technical staff of more than 600 people in eight research departments and fifty research groups, of which twenty-seven are professors, ninety-one are associate professors, and twenty-three are senior engineers. The Institute also has a library, a central laboratory, a workshop, an aquatron, a fish and shrimp mariculture station at Huangdao, and maritime stations at Yantai and Shamen (Amoy). The Institute maintains three research vessels. The library has a collection of 150,000 volumes and about 4,000 Chinese and foreign periodicals, of which about 1,300 are current subscriptions. The Institute edits three journals in Chinese, with English abstracts.

All three institutions are multidisciplinary and responsible for surveying physical environmental data, the status of resources, their distribution and growth patterns in the adjacent sea areas, and the impact of human activities on the marine environment. These data form the scientific basis for marine resource management and ocean services, as well as for ocean engineering projects and marine environmental protection.

Tienjin is the venue of four major institutions: the Institute of Marine Scientific and Technological Information, the National Oceanographic Data Center, the Institute of Ocean Technology, and the Institute of Seawater Desalination and Multipurpose Use.

The Institute of Marine Scientific and Technological Information and, concurrently, the National Oceanographic Data Center, are mainly responsible for the organization and coordination of national work in marine scientific and technological information and oceanographic data; the formulation of management regulations, development programs and technical standards in national marine scientific and technological information and oceanographic data; the establishment and management of a marine literature data base, a marine information data base, and a national oceanographic data base; provision of marine information and data services to users; and participation in relevant international organizations and their activities.

The Institute of Ocean Technology is a national Institute for research concerning marine environmental observation techniques, and it provides technological systems and instrumentation for marine observation, monitoring, and investigations. It also operates the Center of Standards and Meteorology of the State Oceanic Administration and the National Marine Meteorological Station, which are responsible for the work of standardization and meteorology appraisal for all the marine units in China.

The Institute of Seawater Desalination and Multipurpose Use specializes in research on the utilization of seawater resources. It operates, among others, the Division of Seawater Multipurpose Use and the Multipurpose Experimental Base.

Situated in Dalian is the Institute of Marine Environmental Protection, an Institute which is mainly engaged in research on marine pollution and its prevention, as well as on the techniques and methods for

Ocean Station at Xiaomai Island, part of the coastal station network responsible for marine environmental monitoring, surveillance, and forecasting services, as well as environmental protection and law enforcement. Photo: SOA

marine pollution monitoring and environmental management. The Ningbo Oceanography School, which is located at Ningbo, Zheijiang Province, is a base for training intermediate marine technical personnel.

Also located in Beijing are the Marine Environmental Forecasting Center and the China Ocean Press.

The Marine Environmental Forecasting Center is a national organization in charge of issuing forecasts and warnings concerning conditions in the China seas, and conducting research on numerical forecast models for the marine environment.

The China Ocean Press specializes in marine sciences, publishing mainly books and journals on marine sciences and technology. During the 1981–85 period, the books and journals published by the Press won a dozen awards, such as the National Scientific Books of Excellence. About ten percent of the total number of books published are reprints. There is close cooperation with foreign countries. Contracts for cooperative publishing have been signed with firms in Switzerland, Germany, the United Kingdom, Japan, and others. Exchange of books and journals with both domestic and foreign publishers has been actively pursued. Exchange arrangements have been made with 172 marine institutions and organizations in thirty countries.

Main Accomplishments

To give an idea of the work achieved since the 1950s, when the Quingdao Marine Biological Laboratory was first established, what follows is a brief survey of the activities of this institution.

It has carried out comprehensive physical and organic resources investigations. Intensive research has been done in aquaculture, sea farming, and sea ranching—in particular, in the mariculture of algae, mollusks, shrimps, and fish, as well as in the enhancement of fishery resources by releasing juvenile shrimp and fish into the natural environment.

Studies have been carried out on marine biological productivity, on the ecology of the China Seas and adjacent oceans, and on the comprehensive utilization of marine biological resources, especially that of seaweed.

The Institute has also conducted taxonomic and morphological studies of marine plants and animals, and completed a compilation of Chinese fauna and flora—including studies on developmental marine biology, as well as comparative studies on photosynthesis and the evolution of algae.

A great deal of work has been done in physical oceanography, and today there is a deeper and more comprehensive understanding of prevailing current systems, water masses, tides, tidal currents, and waves in the seas around China. Intensive studies have also been made on bottom topography, sedimentation, nutrient salts, and tectonics.

A long-range study of sea circulation (including the Kuroshio Current) has been undertaken, including research on large-scale air-sea interaction and its influence on climate and weather. An important feature in this project was a systematic study of the effects of West Pacific circulation on offshore circulation and the change of climate.

Investigations concerning the sedimentary and tectonic characteristics of the continental shelf of the East China Sea, the Yellow Sea, and the Okinawa Trough (and the mechanism of sediment transport) are being conducted to acquire basic working data for long-term assessment of potential oil and gas resources. Currently, the emphasis is on the geological problems of oil and gas exploration in the Bohai and Yellow Seas.

There have been notable discoveries of new physical marine phenomena, such as the South China Sea Warm Current and the East China Sea Mesoscale Eddy.

In the marine engineering sector, research has been undertaken on the effective prevention of marine boring and fouling on pipes, ships, and harbor installations—as well as on the erosion of metals in seawater and its prevention.

The Institute has developed more than forty instruments for marine scientific research, published more than 1,660 scientific papers and reports, and edited well over twenty atlases. For its achievements, it has received over sixty scientific awards.

The Institute cooperates with about 900 scientists from Britain, Canada, France, Japan, the United States, Germany, New Zealand, Norway, Sweden, the Soviet Union, Thailand, Algeria, Mozambique, the relevant United Nations organizations, and Hong Kong.

Research Vessels, Equipment, and Instrumentation

The State Oceanic Administration operates some forty research vessels with a total tonnage of over 60,000 tons. Seven of these are deep-sea research vessels, capable of conducting research in the oceans around the world.

Two periods, 1965–66 and 1970–77, were devoted to intensive efforts to improve China's marine instrumentation. In recent years, while implementing the policy of exploring exchanges with the outside world and introducing advanced technologies from foreign countries, the State Oceanic Administration has been actively promoting research and development in instrumentation for remote-sensing, seawater desalination and water treatment, as well as submerged floats and buoys. (Hydrometeorological buoys and satellite receivers are partly imported from abroad, partly made in China.) SOA now owns over a thousand pieces of various equipment and instruments. The equipment and instrumentation of ocean measurement and services have also been much improved. A few examples are the Precision Conductivity Ratio Measuring System (Model JDA. 1–1), China's standard equipment with the highest precision for salinity measurement. The SZC Series STD Recorder is an instrument capable of making continuous measurements of seawater temperatures and salinity values at various depths and computerizing the data. Its maximum operational depth is 6,000 meters. It was developed in 1984 for the first expedition to the Southern Ocean, and it proved to be highly successful.

SOA has established a number of laboratories that have reached advanced levels by world standards. Among them, the Marine Controlled Ecology Laboratory conducts experimental research on the dynamic processes of pollutants, such as heavy metals and hydrocarbons, and their impact on marine primary productivity. These projects are currently being carried out jointly by China and Canada. This lab has undertaken the task of international inter-calibration of samples for the intergovernmental Oceanographic Commission of UNESCO. There is also a radiochemical lab which has been used in efforts to set ocean radioactivity standards. Research has been conducted on the radioactivity background in various sea areas.

Of particular interest is the development of seawater desalination technology. Seawater can be desalinated through a distillation process. Fresh water may be a by-product of Ocean Thermal Energy Conversion (OTEC), a process that generates electricity through a turbine driven by the low-pressure steam produced when warm surface water is brought into contact with cold bottom water. The steam subsequently precipitates and the result is distilled fresh water.

Another desalination technology is called "reverse osmosis." Osmotic pressure, generated by the contact between salt water and fresh water (or between more and less saline water), through a membrane, can be utilized for the production of huge quantities of energy, at the mouths of great rivers, around salt domes and icebergs. Technologies to exploit salinity differentials for the production of energy are still on the drawing board. The same principle, however, is already being widely applied, on a rather large scale, to the production of fresh water by forcing the salt content through a membrane.

China started research in this area as early as 1958, and after thirty years of intensive work, has assembled a specialized contingent of over 3,000 people engaged in scientific research on seawater desalination, as well as more than 100 research, design, and information units, and nearly 100 manufacturers. To date, over 300 research projects have passed technical appraisal. Some of these have reached the highest level by international standards, and some have already been commercialized. SOA has made some major breakthroughs in research on electrodialysis and reverse osmosis desalination technologies. The most important and most successful projects include: "research on frequently-rotating electrode electrodialysis"; "formulation and development of technology for ester acid fiber millepore filter membrane"; "4-inch, 8-inch hollow fiber reverse osmosis desalination device; and roll-type hollow fiber reverse osmosis desalination device." Some of the products have approached the quality of the best foreign products of the same type.

With growing population pressures, urbanization, industrialization, and the high cost of energy, the crisis in the use and abuse of water resources has become a serious problem that is causing chain reactions in society, economy, and ecology. The ocean occupies over seventy percent of the earth's surface, and thus its water supply, for all intents and purposes, seems inexhaustible. Seawater desalination is an important means of exploiting and utilizing water resources. It is now becoming a multipurpose applied technology and is seriously being studied by many countries in the world. Small islands, such as Malta, (with which China has a number of cooperative arrangements), already produce as much as sixty percent of their water supply from seawater through reverse osmosis. Here is a new area of marine technology development that may well be of crucial importance for the future.

Oceanographic Investigation

During the past thirty years, China's oceanographic investigation capacity has developed rapidly. A large

Prawn Nursery Pond. Baby
prawns are transferred to the
nursery pond from spawning
tanks. More research is
needed concerning the
spawning of shrimps.
Although shrimps do not
spawn naturally in captivity,
various techniques have been
developed to induce
spawning. Photo: SOA

Artificially-raised prawns.
Prawns are widely cultured in
China, Japan, and all of
Southeast Asia, and—more
recently—in other parts of the
world, as well. Farming
facilities range from the
simplest tidal ponds to the
most sophisticated,
completely controlled
environments in energy- and
capital-intensive tanks. Photo:
SOA

Launching a sea-current buoy
developed by the Institute of
Oceanology. Photo: SOA

R/V *Xiangyanghong* 09. This is a
deep-sea, multi-purpose
vessel equipped with a
number of laboratories for
work in various disciplines.
Photo: SOA

Seawater desalination station
with a daily output of 200
tons of fresh water, built on
Yongxing Island, Xisha, in June
1981. Photo: SOA

number of comprehensive and multidisciplinary investigations of offshore and deep-sea areas have been carried out. China has actively participated in a number of international cooperative investigations. At present, observations are carried out about 3,000 station-times every year, and data for most of China's offshore areas and part of the deep-sea area have been gathered, and a basis has been created for marine scientific research and ocean development and utilization.

Since the seventies, the R/V *Xiangyang* 16 and other research vessels have successfully carried out exploration for manganese nodules in the Central and Northern Pacific. In recognition of this exploratory work, China has recently been awarded the status of "Pioneer Investor" by the Preparatory Commission for the International Seabed Authority and for the International Tribunal for the Law of the Sea. China is the fifth State to be awarded this status, after India, France, Japan, and the Soviet Union.

In 1978, R/V *Shijian* and R/V *Xiangyanghong* 09 participated in the First GARP (Global Atmospheric Research Project) experiment, coordinated by the Intergovernmental Oceanographic Commission of UNESCO.

Between 1980 and 1986, a number of national projects for the exploration, development, and utilization of coastal zone and tidal wetland resources were carried out. All these projects were broadly interdisciplinary, including economic and environmental aspects. The series was successfully completed in 1986.

During the same period, two expeditions to the Antarctic and the Southern Ocean were undertaken, and a scientific research base was established on the Antarctic continent. This entitled China to become a Consultative Party to the Antarctic Treaty.

Also during this period, a number of joint expeditions were organized between SOA and the United States, Japan, and the Federal Republic of Germany.

Services

Services such as weather forecasting, information, etc., constitute an important part of SOA's activities. Since 1982, the Administration has been FAXing ocean wave charts daily to users both at home and abroad. Ships sailing in the Bohai Sea, Huanghai Sea, the East and South China Seas, as well as in the western Northwest Pacific, may receive ocean wave facsimile charts which,

with their clear images, help greatly to reduce the chances of accidents caused by rough seas. Facsimile transmissions of other oceanographic elements, such as storm surges, sea ice, currents, etc., will be introduced step-by-step in the future. Since 1986, forecasts for sea waves, sea temperatures, and storm surges have been broadcast to the whole country through the Central Broadcasting Station and through Television. Moreover, fixed-point weather and ocean forecasting services are provided to the various departments for oil exploration and exploitation, and optimum ship routing services, as well as special routing and offshore operating services, are provided to ships sailing in the Pacific and the Indian Ocean.

In recent years, SOA has developed a comprehensive basis for marine information technologies, consisting of marine data, marine information and marine services, and scientific and technological services for public welfare. Services in marine data and information are also made available to the departments for scientific and industrial research and decision-making, and are well received by users both at home and abroad.

Law Enforcement

Together with other interested departments, the State Oceanic Administration has established a National Marine Monitoring Network to safeguard rights and interests in the ocean, and to monitor and enforce compliance with standards and regulations. To start with, a volume entitled *Provisional Standards for Marine Pollution Survey* was published, the results obtained by various laboratories for national marine pollution monitoring were intercalibrated, national standards for seawater quality were formulated, and a number of laws were enacted, such as the "Marine Environmental Protection Law of the People's Republic of China," which covers offshore oil exploration and exploitation, "Regulations of the PRC concerning Dumping of Wastes at Sea," "Coastal Zone Management Law," etc. A surface oil identification system including field investigation, sample collection, storage, transportation, and preservation has been established using various methods for analysis and identification. In recent years, this system has been used to handle over twenty important controversial oil spill cases involving foreign vessels. To enhance law-enforcement management at sea, a remote-sensing monitoring and surveillance system was established in 1984, on a frequent but nonperiodic basis.

The "Great Wall" Antarctic Station. During the 1980s, two expeditions to the Antarctic and the Southern Ocean were undertaken, and this scientific research base was established on the Antarctic continent. Photo: SOA

International Cooperation

SOA has actively participated in a number of important international activities, strengthened contacts with other countries, and joined a number of agencies of the United Nations system of intergovernmental organizations as well as nongovernmental organizations. In 1973, China became a member of the Intergovernmental Oceanographic Commission of UNESCO. Since 1983, China has been sending a Delegation to the sessions of the Preparatory Commission for the International Seabed Authority and for the International Tribunal for the Law of the Sea, which are held every year in Kingston, Jamaica, and in New York. In 1984, SOA and IOC/UNESCO jointly organized a training course in Tienjin on tidal observation and data processing. Lectures and instruction in practical operation were given to trainees from twelve countries in Asia, Africa, and Latin America. Another interna-

Gulf scallops that have been
artificially cultured. Intensive
research has been undertaken
in the mariculture of mollusks
at the Quingdao Institute of
Oceanology. Photo: SOA

tional training program, in the management and conservation of marine resources, was organized by SOA
in cooperation with the International Ocean Institute
of Malta. A similar course is being organized for 1992.

In 1986, China became a member of the Aquatic
Science and Fisheries Information System (ASFIS) es-
tablished by the Food and Agriculture Organization
(FAO) of the United Nations.

Conclusion

This is just a sampling of China's international activities in an area where national, regional, and international issues are inseparably linked. China has made
an enormous effort to build a national infrastructure
for marine sciences and technology, and to develop the
human resources required for ocean management.
Without such an infrastructure, international cooperation remains illusory. At the same time, the building of
a national infrastructure often requires international
cooperation and assistance—China has tried to move
on both levels simultaneously.

China has also recognized the interdisciplinary
nature of ocean management. As the 1982 United Nations Convention on the Law of the Sea states, "the
problems of ocean space are closely interrelated and
need to be considered as a whole." This means that
ocean management requires integrated policy making
and planning in which a number of government departments need to be involved. China has tried to
reflect this recognition in its governmental structure
and in the structure of the State Oceanic
Administration.

The potential of the oceans' contribution to
China's food, water, mineral, and energy supply during
the twenty-first century is impressive, to say the least.
The Chinese people, the Chinese scientific and industrial community, and the Chinese Government have
been working, and will continue to work toward utilizing the final decade of the twentieth century as a
preparatory period for realizing this potential.

9.

The Ocean
Research Institute,
Tokyo

Takahisa Nemoto

The Ocean Research Institute.
A part of the University of
Tokyo, ORI is the core
institution for the marine
sciences in Japan. Photo: ORI

Introduction

Japan depends on the sea for its very survival. Its landmass, of 377,643 square kilometers, consists of eighty percent uninhabitable rugged mountains and only twenty percent habitable land, with a density of population that is several times higher than that of the large Western European countries (Germany, the U.K., Italy, and France)—and even more than three times higher than that of the Western European countries with the highest population density (the Netherlands and Belgium). Most of the congested cities in Japan are located along the coasts, which are very long. They cover over 32,000 kilometers, including the shore lines of more than three thousand Japanese Islands. For each square mile of land surface there are nine miles of coast line. No place in Japan is more than seventy-five miles from the sea, which has made Japan a naturally seafaring country and a leading fishing nation. Sardine, salmon and seabream, yellowtail, horse mackerel, tuna, trout, shark, flying fish, mullet, smelt, and cod are abundant in the huge Economic Zone Japan acquired as a signatory of the United Nations Convention on the Law of the Sea: This covers an area of 4,510,000 square kilometers, which is twelve times larger than the landmass. A great number of species of clams, oysters, abalone, and sea snails are used for food; mussels, oysters and shrimp, seabream, yellowtail, salmon, and eel are farmed commercially—in aquaculture, Japan has become a leader.

At this time, there are 1,085 commercial ports in Japan, which means, every thirty kilometers of the coastline has a commercial port, and there are about 2,800 fishery ports.

Harbor construction, reclamation, and coastal protection are important industrial activities in Japan. Japan is also a leader in the construction of artificial islands, in ship building, and in marine industrial technology. The Third United Nations Conference on the Law of the Sea recognized Japan, together with France, India, and the Soviet Union as "Pioneer Investors" in deep seabed mining, and granted it exclusive tenure over a mine site in the Northern-Central Pacific.

A number of Government departments are actively involved in ocean development. Thus, one of the major objectives of the Ministry of Transportation for the utilization of ocean space is an offshore man-made island project. The Ministry describes the purpose of the project as follows:

> In order for Japan, with its limited land area, to achieve balanced development, it is a vitally important task to push forward ocean space development and utilization based on a long-term perspective, while promoting effective and efficient utilization of the existing land area. For this purpose, research and study efforts are necessary, because Japan's stable development, as well as the creation of an affluent and comfortable society are dependent on the expansion of the country's land and space.

The Ministry of Construction is charged with maintaining "a safe and comfortable" coastal zone to meet the increasing demand for marine recreational activities. The Ministry of Agriculture, Forestry, and Fisheries is responsible for securing a stable supply of marine products and helping regional development by introducing new fisheries technology in response to the need for development of marine resources in the coastal zone. The Ministry has been engaged in a number of major plans to achieve this main goal: These plans include:

... The Marine Kombinat Plan—the development of large-scale urban fishery centers and the expansion and stabilization of offshore ocean resources.
... The Marine Village Plan—the development of the fishery industry sector engaged in hatchery and cultivation activities, and the improvement of the living environment of fishing villages.
... The Marine Technology Plan—research and development centered on introducing high technology to the fishery industry.
... The Marine Culture Plan—the preservation of marine cultures and the fishing environment.

The Ministry of International Trade and Industry has been carrying out a Marine Community Policy Plan to integrate various types of technology for the efficient exploitation of ocean space.

There is a Ministry for the Environment, whose

Pop-Up System. This system, containing oceanographic instrumentation, such as current meters, temperature, salinity, and density measuring instruments, etc., is dropped from a ship. Attached to it is a weight, or a series of weights, which are deployed when a radio signal is given, to lower it to a determined depth, or series of depths. After its recordings have been completed, buoyancy equipment, also attached to the system, makes the system "pop up" to the water surface. It then sends a radio signal to the mother ship so that it may be picked up. Photo: ORI

responsibilities include the construction of facilities such as breakwaters, promenades, or artificial seashores.

Finally, there is a Science and Technology Agency that has adopted a comprehensive "Aqua Marine Plan" with three major components:

... Promotion and diffusion of marine technology to regional communities to facilitate the comprehensive use of ocean space.
... Implementation of policy measures for responding to local needs and developing marine science and technology.
... Improvement of the ocean development potential of regional communities to strengthen the foundation of ocean development in Japan.

The Ocean Research Institute: Structure and Functions

The rational use of ocean space and resources rests on effective marine scientific research. The more intensive this use, the greater its dependence on high technology—and high technology, in turn, depends on science. Without an understanding of the physics and chemistry of the oceans, the workings of currents, tides, and waves, without the ability to monitor changes wrought by human activity on the nature of the ocean, long-term gains turn into short-term, immediate costs, and economic benefits are wiped out by ecological losses.

Japan is keenly aware of this equation, and its stress on the importance of marine scientific research is commensurate.

The Ocean Research Institute (ORI), which is part of the University of Tokyo system, was founded in April 1962 by an Act of the Japanese Diet. During the preceding years, many Japanese marine scientists and scientific societies had emphasized the need for establishing a new oceanographic institution for basic and comprehensive research in ocean sciences.

In April 1958, the Science Council of Japan passed a resolution at its 26th General Assembly, recommending the establishment of such an institute to the Japanese Government. Four years later, the Government decided to appropriate the necessary funds for use by the University of Tokyo. Since then, the Ocean Research Institute has functioned as the core institution for marine sciences in Japan.

The Institute presently has fifteen research divisions with facilities on the Nakano Campus. A shore laboratory, the Otsuchi Marine Research Center, is the equivalent of an additional, sixteenth division. It is located in Otsuchi, Iwate prefecture.

The Institute operates two research vessels, the *Tansei-Maru* and *Hakuho-Maru*, under the auspices of the Ministry of Education, Science, and Culture (Monbusho). These two ships, as well as the Otsuchi Marine Research Center, are accessible to Japanese marine scientists throughout the country and are often utilized in the implementation of international projects. The Institute also supports more than fifty graduate students at the Master's and Ph.D. levels, and other research students in physical, chemical, and biological oceanography, meteorology, geology and geophysics, marine biology and fisheries. About 7,000 to 8,000 researchers, including scientists and students from all over the world and a faculty of about sixty ORI scientists, are involved every year in ORI's research efforts. The Institute participates in numerous international projects, in cooperation with the United States, France, the U.K., the U.S.S.R., as well as UNESCO, IOC. The R/V *Hakuho-Maru* has conducted many cruises for the Training and Education Program (TEMA), under IOC sponsorship.

The University of Tokyo, to which ORI is attached, is one of the largest in the Japanese university system. That system consists of ninety-six national universities, supported by the Monbusho. In 1986, Monbusho made funding of 5,167 billion yen available to its universities and institutions for research and education. Additional funds are allocated for specific projects, which are carefully screened before approval and funding. ORI's budget is steadily increasing.

Ocean science is one of nine major fields of research. It covers all branches of oceanography and the manifold interactions between mankind and the marine environment: the exploitation of living resources; the protection of the marine environment, as well as the nonliving resources in the water column and the deep ocean floor.

Each of the fifteen scientific divisions is headed by one Professor, assisted by one Associate Professor, two Research Associates and two or three technicians. The Otsuchi Marine Research Center, which has the status of a scientific division and a similar structure, is headed by a Professor who works with four Research Associates.

The fifteen divisions and the Otsuchi Center conduct their work in five main areas: marine physics, which covers both physical oceanography and marine meteororology; marine chemistry, including both inorganic chemistry and marine bochemistry; marine

Marine Research Center at
Otsuchi. The areas of
investigation here cover
physical oceanography and
marine meteorology, marine
chemistry, geology and
geophysics, as well as biology
and fisheries science. Photo:
ORI

Acropora (coral). Coral reefs are
of vital importance to the
stabilization of coastlines, as
fish habitats, and for the
protection of bio-diversity.
Photo: courtesy Jun Oui

Acropora (coral). Many of the
splendid coral reefs in the
seas surrounding Japan are
dying due to the heavy
pollution of the coastal
waters. Photo: courtesy Jun
Oui

The research vessel *Tansei Maru*, operated by ORI under the auspices of the Ministry of Education, Science, and Culture. Photo: ORI

geology and geophysics, including submarine sedimentation and ocean floor geotectonics; marine biology, which covers the physiology of marine organisms as well as marine ecology, planktology and microbiology; and, finally, fisheries sciences, including population dynamics, biology of fisheries resources, fisheries oceanography, and systems analysis of fisheries resources.

Oceanographic research in Japan covers a broad geographic area, ranging from tropical waters to boreal seas, and from intertidal zones to hadal depths. A number of important projects have been carried out in the Sea of Japan, the Philippine Sea, the East China Sea, and the North Pacific. The R/V *Hakuho-Maru* completed many cruises in the Arctic as well as the Antarctic Oceans to participate in international programs.

All divisions have made significant contributions to oceanographic research. Perhaps one might mention in particular Division 15, working on a systems-analytical approach to the study of fisheries resources and employing the most advanced high technology in this work.

Yayoi ship of the Otsuchi
Marine Research Center, which
is used in nearshore
oceanographic research.
Photo: ORI

The objective of this division is to develop techniques for the assessment of the abundance of fish and micronekton and to observe ecological features applying image-processing and artificial intelligence techniques. This includes also studies on the mechanisms influencing stock size. It involves:

... The application of hydro-acoustic techniques to stock assessment and behavioral studies.
... Development of an underwater observation system combining deep-sea camera and hydro-acoustic detection for the measurement of size, observation of behavior and identification of fish and micronekton.
... Analysis of schooling behavior of fish. How do fish aggregate and maintain the integrity of the school? How does schooling benefit the individuals? These questions have been studied both through laboratory experiments and by computer simulations.
... Studies of the life history and ecology of certain fish.
... Studies of the early life history of fish, including the development of fish eggs and larvae; the feeding

habits of fish larvae, and the causes and mechanisms of mortality of fish eggs and larvae.
... Studies on the efficiency of gear on marine animals.
... Studies on the development of a method applying image processing and artificial intelligence techniques.
... Analysis of fishing conditions using satellite information.

Some of the major research projects of the other divisions are briefly described on the following pages.

... The Mysterious Eel

Professor Okiyama and Dr. Tsukamoto are working on comparative studies on fish migration, including ecological, physiological, oceanographic, and behavioral aspects. Anadromous fish, such as salmon; catadromous species such as eel, as well as other oceanodromous fishes have been studied. Anadromous fish breed in fresh water, the young then migrate to the oceans, where they spend most of their lives, before returning to their native rivers, to breed and die. Catadromous fish breed in the oceans, migrate to fresh water, and return to the oceans to breed and die.

Atlantic eels have been studied for many years. For unknown reasons, they travel from the rivers of Europe and the Americas to meet in the Sargasso Sea, near Bermuda. There they spawn in the thick floating weeds of this current-bounded sea—itself a mystery to science and fraught with lore and legend. Then the young larvae depart—those of European parentage take up the long journey, riding the Gulf Stream and the North Atlantic Current for three years to the coast of Europe. The American elvers have a less strenuous journey to complete. It takes them only six months to a year along the Gulf Stream to make their way into the American rivers from which their parents came. After ten years, they all embark on the long hazardous return voyage to the faraway Sargasso Sea. What signal initiates their odyssey, what compass guides them, no one knows.

Even less is known about the Pacific eel, of which there are some fifteen species, widely distributed between Australia, New Guinea, Indonesia, the Philippines, Taiwan, China, Korea, and Japan. Biological information on the early life history of these Pacific eels is quite inadequate. The largest number of slender-headed, transparent juvenile "glass eels" or *leptocephali* ever taken by researchers was seventy-eight,

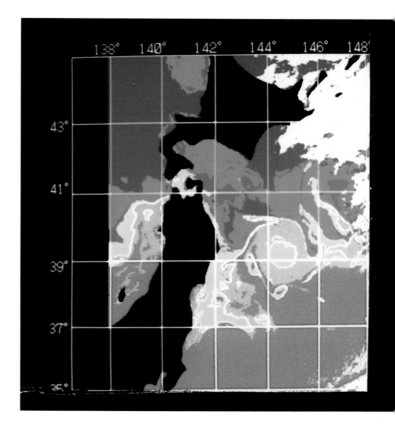

Satellite view of warm core ring. Photo: ORI

and they already measured 34 mm, while more than twenty thousand of the Atlantic eels had been obtained, and the smallest measured only 3.9 mm and seemed to be a newly hatched larva, not more than one week old.

The R/V *Hakuho-Maru* has carried out four survey cruises since 1973 in the Western North Pacific to determine the breeding grounds and migration mechanisms of the Japanese eel, *Anguilla japonica*. Seventy-six *leptocephali* of this species were obtained, ranging from 34 to 60 mm in size. ORI has developed a new technique to determine their age during the latest survey cruise in 1986, and new aspects of eel research are progressing rapidly. The age and birth date of *leptocephali* are now determined by studying the daily growth increments of their otoliths, or "ear stones," i.e., the minute calcarious particles found in the inner ear of certain vertebrates. These analyses showed that the *leptocephali* of about 40 to 50 mm taken east of Taiwan and the Philippines in September were about 70 to 90 days old and hatched in June or July. It was also determined that *leptocephali* grow linearly at 0.46 mm per day and reach a maximum size of approximately 60 mm, 125 days after hatching, just before metamorphosing to the elver stage.

ORI scientists also studied the feeding mechanisms of these organisms, and their vertical distribution in the ocean. They found them to be most abundant in the layer just above the thermocline, at a depth of 75 to 100 meters. They were quite scarce in the deeper layers, suggesting that distribution of *leptocephali* might be restricted by the thermocline. It seems the migration period of the Japanese eel is much shorter than that of the Atlantic eels.

In addition to the field surveys, laboratory experiments are being carried out in order to compare and combine with the results from the field. Artificially fertilized eggs and the resulting larvae of the Japanese eel and the pike eel are used for the study of the development of various organs and the control of *leptocephali* metamorphosis by endocrine and environmental factors. The ultimate goal of this research is the better understanding of the mechanism and processes of migration, and the levels of recruitment in the eel stocks, as well as the intrinsic and extrinsic factors governing them.

The eel is an important food fish in Japan. It is intensively farmed, in totally controlled environments. However, it does not breed in captivity, and farming thus depends on the availability of elvers captured in estuaries and river mouths.

The research carried out by ORI should contribute to increasing the volume of eel fishing from the wild, as well as the further improvement of eel farming.

. . . Microbial Ecology

This is the work of Professor Simidu's division. Some projects are carried out in cooperation with other divisions.

The marine environment abounds in bacteria. Even in the most nutrient-poor areas, remote from land, we can count ten thousand or more bacteria per milliliter of water. However, the counting of living marine bacteria is still a problem, since there is no secure method of discriminating living from nonliving bacteria. A technique, known as the direct viable count (DVC) method, has been developed to solve this problem. The basic principle of this method is to incubate the bacteria in a water sample, preventing fission (bacteria multiply by fission). The resulting elongated, plump cells of the bacteria in the sample allow the easy discrimination and counting of actually growing bacteria.

Along with the effort of counting living marine bacteria, an attempt has been made to determine their biomass and growth rate, using a particle counter perfected by the Elzone Company, Ltd. This work was carried out in collaboration with the Marine Biochemistry Division. The results show that previous measurements vastly underestimated the bacterial biomass because they were based on inadequate conversion factors between the size of individual bacteria and their biomass as a whole. Consequently, it now can be assumed that a much higher percentage of the organic carbon produced by phytoplankton in a water body is converted into bacterial biomass than had been assumed by earlier researchers. This, in turn, may have an impact on the so-called "greenhouse effect," which may be counterbalanced by the oceans' capacity of carbon dioxide absorption.

The taxonomy of marine bacteria is not yet fully developed. However, there are two groups that have been investigated most extensively: the members of the family *Vibrionaceae* and the genus *Alteromonas*. The bacteria of both groups actively participate in the circulation of organic and inorganic matter in the sea. A comparative study of marine *Vibrionaceae* has been carried out at the Marine Microbiology Division. A total of 405 strains, which had been isolated from the Indian Ocean, the South and East China Sea, the West Pacific

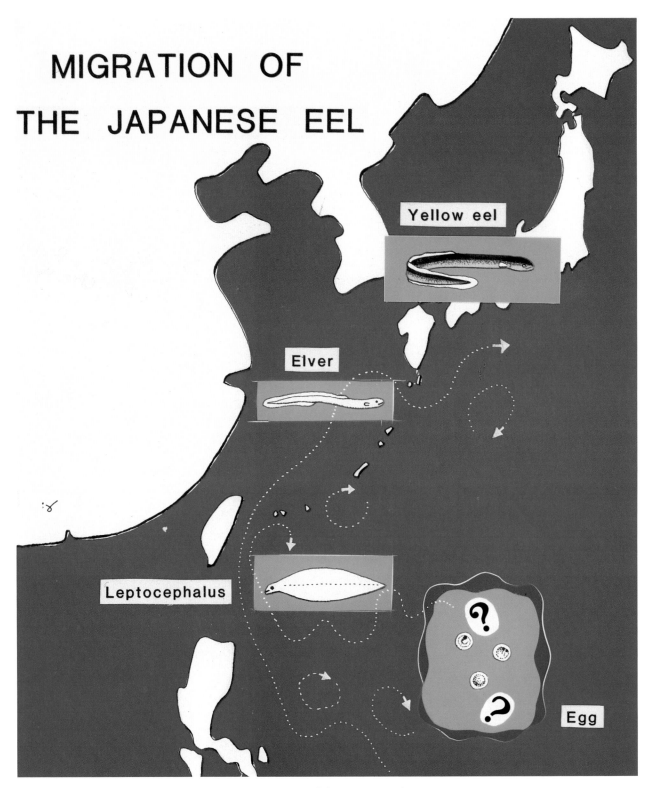

Migration of the Japanese eel.
Drawing by Jiro Komatsu,
August 1987. Photo: ORI

Ocean and coastal areas of Japan, were studied. The results showed that most groups have their particular ecological niche. For instance, each group of seawater vibrios inhabited a particular water layer of limited depth range, even though strains of the group were isolated from sampling locations spread over a wide area ranging from the Indian Ocean to the Japanese coast. Various vibrio groups showed remarkable differences in their physiological and biochemical activities.

Many seepage areas, warm and cold vents or "smokers," have been found around the islands of Japan. These are areas where the seawater penetrates through fissures on the ocean floor into the interior of the earth's mantle, where the water is heated and absorbs a variety of metals. It then returns to the surface of the ocean floor like a submarine geyser; the contact with the cold water precipitates the metals which accumulate in the "smokers," usually at a depth of about 2,000 to 3,000 meters. Often, these formations are associated with dense colonies of animals and microorganisms similar to those found at the hydrothermal vents near the Galapagos Islands and the East Pacific Rise: giant worms, living in clusters, as well as crabs and giant clams. In the absence of light and, therefore, of photosynthesis, these strange animals get their nourishment through chemosynthesis, that is, the activity of the bacteria which utilize the inorganic sulphur compounds, hydrogen, and methane to produce biomass. This process has been studied in collaboration with the Marine Ecology Division and the Marine Inorganic Chemistry Division.

Another aspect of the deep-sea microbiology is the study of barotolerant and barophilic bacteria, that is, bacteria which tolerate—or even prefer—the high pressure inherent to the deep-sea environment. Some of these bacteria cannot survive reduced pressure, and therefore water samples for the study of such barophiles must be kept at the *in situ* pressure.

Increasing demand for fish of high quality has resulted in the rapid development of aquaculture systems in Japan, as well as in many other Asian countries. Environmental deterioration is an important limiting factor. To control this factor, the ecosystem of aquaculture facilities, the composition of microbial populations in the pond (ranging from bacteria to protozoa), and the succession of populations accompanying the changing environmental conditions, have been studied in various types of aquaculture facilities in Japan, Thailand, and Taiwan. Some species of bacteria and protozoa were found to have the ability of keeping or restoring the desirable water quality of ponds.

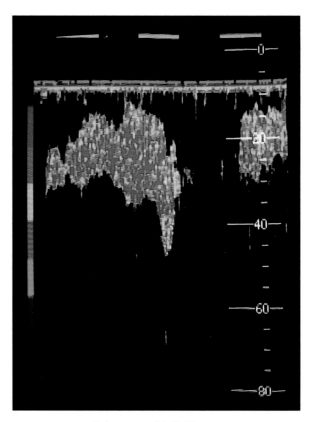

Echo trace of krill. The application of hydro-acoustic techniques to stock assessment and behavioral studies is part of a systems-analytical study of fisheries resources. Photo: ORI

...Marine Pharmacology: Potential Boon for Tropical Countries

Many marine micro-organisms produce biologically active substances: enzymes, enzyme inhibitors, and compounds having antibiotic, antitumor, anti-leukemic, and other pharmacological values. New types of enzymes have been discovered at the Marine Microbiology Division: such as halophilic proteases, that is, enzymes, thriving in high-salinities, which catalyze the hydrolytic breakdown of proteins; psychrophilic proteases, that is, enzymes thriving at low-temperatures, as well as enzymes that inhibit the development of these agents. Other enzymes have been found that convert starch into sugar (*amylase*). In the search for protease inhibitors, nearly 3,000 strains of marine bacteria from various environments were screened, and only three strains of inhibitors were found. These were members of the genus *Alteromonas*. The inhibitors were purified, and in one case their amino acid sequence was determined. It was found that the sequence of the inhibitor of marine bacteria is distinct from all other inhibitors found in land organisms.

Production of animal and plant toxins by marine bacteria is another topic of the current research in marine microbiology.

So far, nearly 3,000 pharmaceutically active substances were isolated from a vast number of marine animals and plants. Among them are many toxins such as tetrodotoxin (pufferfish toxin), paralytic shellfish toxins, palytoxin, and others. Recent studies indicate that they are the products of bacteria that are associated with the animals and plants that carry these toxins, and thus they can be produced in the lab, without the associated animals.

The isolation of strains of bacteria capable of determined biological activities, and genetic engineering to enhance these capabilities, is going to be of vital importance, not only for the pharmaceutical industry, but for a number of other industries as well: replacing chemical and mechanical processes with biological processes, e.g., for the clean-up of oil spills or the extraction of metals from ores, through bacterial systems.

...The Fluctuating Sardine Population

SARP is the acronym for Sardine-Anchovy Recruitment Project. This is a global project carried out under the auspices of the Intergovernmental Oceanographic Commission (IOC) of UNESCO. In Japan it is conducted by Professor Ishii, Dr. Kawaguchi, and their colleagues. This work is of great importance, not only scientifically but also economically, since the Japanese sardine fishery yields an annual harvest of two to four million tons, which amounts to twenty to forty percent of the total fisheries catch in Japan. In the sixties and seventies, however, the stock size was very small—the catch was as low as ten thousand to one hundred thousand tons per year. The fluctuating nature of the clupeid family, which includes herrings and menhadens as well as sardines and anchovies, is well known throughout the world. The collapse of the sardine fishery in California, as well as of the Peruvian anchoveta, the Chilean sardine, and the Scandinavian herring, are famous examples. Some of these losses have caused enormous economic hardship. The drastic fluctuation of the Japanese sardine populations has been known to be periodic, and about six prosperous periods have been recorded since the seventeenth century, suggesting that such fluctuations existed even when the fishing pressure was negligible.

Now, new techniques are being applied to the study of sardine, anchovy, and related clupeid fishes: an egg production method, birth date analysis of juveniles, and the anatomical study of microscopic structures to evaluate the survival rate of larvae. At present, SARP activities are mainly restricted to upwelling areas characterized by low species diversity, such as the eastern part of the Pacific and Atlantic. In the western boundary current regions this work appears to be much more difficult because of the effect of strong currents, such as the Kuroshio and the Gulf Stream, which cause large fluctuations in the survival rate of eggs and larvae, and also because of the much more complicated food and prey-predator interactions in the subtropical and tropical communities.

In this context, ORI has conducted studies in the Kuroshio region, covering not only sardines but also anchovies, round herring, and sand eels. These studies include: the early life history of the Japanese sand-eel, based on otolith aging of larvae and juveniles; the development of sense organs and behavior in marine fish larvae and juveniles; comparisons of nutritional conditions of Japanese sardine larvae between offshore and inshore regions of the Kuroshio axis on the basis of growth rate, etc.

Now, adult anchovy have been successfully reared in captivity at ORI, and eggs are available for experimentation every day from August to November. The establishment of sardine rearing techniques will increase the understanding of recruitment processes

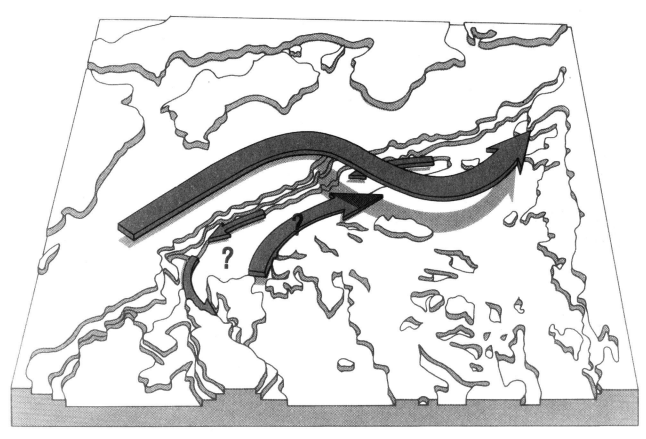

Drawing of sea current. ORI has conducted many studies in the Kuroshio region, and investigated the impact of the current on the nutritional conditions of Japanese sardine larvae. Courtesy: ORI

and the fluctuation of stock size of Japanese sardines.

What we have presented in these pages is but a random sampling of ORI activities, highlighting the advanced technology and methodology, the scientific-frontier position, and the economic importance of the Institute's work. Important ecological studies and research in marine chemistry, such as the research on the hydrothermal activity over the Loihi Seamount, have been omitted for lack of space. This latter project was of particular interest because of the use of a shipboard gas chromatographic system that gave a fantastic picture of dissolved methane distribution over the summit of the Seamount. It became clear that the summit is completely covered with hydrothermal plumes that have an abnormally high concentration of methane, possibly originating from the eruption of hydrothermal fluids somewhere on the summit.

Japan is dependent on the ocean, and the rational use of the ocean is dependent on marine scientific research. An intensification of ecological studies and integrated coastal management studies will be needed if development and conservation goals in Japan are to reinforce rather than obstruct each other.

10.

The Peruvian Marine Institute, IMARPE, Callao

Geoffrey L. Kesteven

The Peruvian Marine Institute,
IMARPE, established in 1964.
Photo: IMARPE

P eru is a country of large dimensions and dramatic contrasts. A visitor who ascends by car or train from Lima to Huancayo quickly becomes aware of these characteristics—and, if he forgets the difference in altitude between the two cities and the difference in the air, he soon discovers one of the effects of these contrasts in his own breathing.

Geographically, the country is made up of four strikingly different provinces: the Amazonian, on the eastern side of the Andes, is hot and wet; the Andean, high and imposing, is often bitterly cold and its air is thin; the coastal plain, most of it a sand desert with little or no rainfall, is crossed by rivers bringing water from the Andes to nourish sugar cane and other crops and to support cities; and, lastly, the maritime province—the Exclusive Economic Zone, claimed under Part V of the United Nations Convention on the Law of the Sea—is large (its area close to that of the other three put together), and its weather strangely capricious.

Each of these provinces has a character all its own, from its geological features, through climate and flora and fauna, to the people and their culture. The ethnic differences are such that a Peruvian once, in conversation, attributed the country's political problems to, as he claimed, the fact that its people were not one, but three nations; he went on to describe the attributes of each: the easygoing easterners, the irascible and suspicious people of the Andes, and the go-getting people of the coastal plain.

Having no residents, the maritime province cannot exhibit ethnic individuality, but it has its own special characteristics: an extremely narrow continental shelf, a continental slope of vertiginous steepness down to a great depth, a complex current system, of which the Peruvian (formerly Humboldt) current is only part, and amazingly abundant marine life. This is the province with which, obviously, IMARPE is concerned.

Anchoveta, Cormorants, Guano

Through all human history, whether of recorded time or the much greater unrecorded time, those who lived at the edge of the sea have taken food from its beaches and rocky foreshore. The inhabitants of what we now know as Peru have been no exception. The massive *Historia Maritima del Peru* cites ample evidence of Peruvians collecting clams and mussels at the edge of the sea in the distant past, and of the many influences exerted by the sea on Peruvian life and culture. This *Maritime History*, of seven volumes (three of them double) that was prepared under the direction of a Commission appointed for the purpose by the Peruvian Naval Center and Naval Museum, gives a detailed account of a close nexus of sea, land, and mankind, and speaks with conviction of "The Sea: a Great Personage." While it fully acknowledges the dangers confronting those who venture to navigate the sea, it describes the sea as generous, "providing man, from the astonishing richness of its waters, with an exceptional wealth which has always been the support of man, from the dawn of time until now."

From being only collectors of *mariscos*—clams, mussels, other mollusks, and shrimps—Peruvians became sea-fishermen and sailors. They developed all kinds of fishing gear (harpoons, hooks, traps, nets) and platforms from which to work. They built rafts (balsas) of reeds, skins, and balsam wood, and thus could exploit resources beyond the beach. The balsa was developed from a simple support for one man into a craft capable of transporting people and cargo, furnished with a tiller and sail. On such a craft, the Inca Tupac Yupanqui is said to have taken a force of some thousands of men to "distant islands" and returned bearing valuable "things"—whatever doubts there may be about the accuracy of the reports of that exploit, there can be none about the reports of coastal traffic on these crafts, carrying people and goods, some time before the arrival of the Spaniards.

By the time of the Incas, the people of Peru were making extensive use of their fishery resources, capturing a wide variety of species of fish and many mollusks. "And when the Inca wished to eat fresh sea fish, there being seventy to eighty leagues from the coast to Cuzco, it was brought to him alive and wriggling . . ." Prescott described how runners, or *chasquis*, each running some short distance as in a relay race, carried fish up to Cuzco within a span of twenty-four hours.

Above: Panama Horse Conch
in the waters off the
Galapagos Islands. Photo:
© 1991 Carl Roessler

Below: Swarming Fish in the
waters off the Galapagos
Islands. Photo © 1991 Carl
Roessler

But the sea yielded more than immediately consumable food; in particular, it provided, from the anchoveta via the sea birds, guano which was of such importance that Gonzalo de Raparaz could write about "the guano civilization," arguing that "the humble guano constituted in the remote past of the Peruvian people a factor of enormous importance, which exercised a decisive influence in the development of the coastal civilization." And it is reported that the system of exploitation, transport, distribution, and application of guano, by which the coastal farmers maintained indefinitely the fertility of their oases, reached a high level of efficiency under the Inca administration.

The Spanish invasion of Peru, in 1532, was in pursuit of gold and silver. Over the years, the invaders carried away very great quantities of these metals; but in time they turned their attention to other resources, to timber and other materials. The export of raw materials continued after independence, and notable among these materials was guano. The exploitation of this material for export began in 1841 and continued for more than a century. During much of that time the exploitation was effected by the *Compania Administradora del Guano*, an organization that was well aware of the ecosystematic relations between seawater, plankton, anchoveta, birds, and guano. For many years the Company kept on its payroll a German expatriate, Erwin Schweigger, who carried out oceanographic research with a view to giving an account of the production of guano by the birds. In addition, in relatively recent times, the Company supported a study of the bird populations, with frequent aerial census of the population of each nesting site.

The guano deposits, such as they were in the time of the Incas, were of very great antiquity, almost geological formations, and as such have been studied by scientists—notably by Evelyn Hutchinson who published a large monograph on the guano deposits of the world. Studies of cores taken through Peruvian deposits showed their great age, and the great variations that had taken place in the rate of deposition, there being evidence of periods in which relatively little dung had been deposited. This evidence was consistent with the observation of bird mortalities: at times the beaches of the Peruvian coast were densely littered for hundreds of miles with the carcasses of guano birds. With such a great mortality rate, the amount of dung produced must have been greatly reduced. The mortalities occurred at times when the coastal waters were warmer and the quantities of anchoveta were smaller: the combination of these conditions was given the name of El Niño because, so it is said, it

occurred at about Christmas, in celebration of the birth of the Christ child.

Development of a Fishery

Knowing, as it did, that its raw material came from the anchoveta via the guano birds, the Guano Company cannot have looked with benevolent eyes on proposals to develop a commercial fishery based on the anchovy stocks. Indeed, there is good reason to believe, even if there is no evidence to prove it, that the Company resisted such proposals. With very good reason: if the anchovy is caught by man in substantial quantities, there will be less on which the birds could feed, and hence less guano. And in the event, so it proved: the bird population before the fishery started was estimated to be in the order of eighteen million; by 1973 the number had fallen to something like two million.

Nevertheless, by the 1950s, the technology of reducing fish to meal and oil was so advanced, and the demand for fish meal had become so strong (offering the prospect of great rewards), that the demand for fishery development could not be resisted. Thus, the anchoveta still had to be exported, not as food for direct human consumption, nor as guano, but as fish meal.

By 1956, the anchoveta fishery was well launched and it was to make Peru, in about ten years, country number one in quantity of sea fish caught. Both domestic and foreign capital was available for this so promising development. Boats were built in large numbers, with main motors from the United States, the United Kingdom, and Scandinavia; with immense purse seine nets, similarly imported; and with powerful pumps to suck the fish from the net-encircled school into the holds of the boats. Onshore, large processing plants were constructed and equipped with mills, filters, driers, and oil separators, also imported, and with powerful pumps to draw the fish from the hold of the boat into the storage bins.

Creation of an Institute

That is where IMARPE came in.

In the 1950s, the General Staff of [the Peruvian] Navy, gathering suggestions from its personnel, proposed to the government the creation of a Council for Hydrobiological Research whose principal mission would be "to coordinate and intensify" studies of the entire ecosystem of which the anchoveta is a vital component. As early as 1945, the Peruvian Government had decreed (as had Chile), its intention to assume sovereignty over the seas as far as 200 nautical miles from its shores. Peru based its claim, at least in part, on a description of the marine ecosystems of those waters and of the relationship between the landmass and the waters. That decree was converted into the Santiago Declaration in 1952, by Peru, Chile, and Ecuador. The General Staff of the Peruvian Navy undoubtedly played an important role in these moves, and was aware of the technical problems that would be encountered in defending the Santiago Declaration before the international community. But Peru, at that time, did not have the scientific specialists who could study the marine ecosystems and present the evidence to support such a national claim. Moreover, the drive to develop fisheries was then under way.

Therefore, in 1958, the Government of Peru sought and obtained from the Specialized Agencies of the United Nations—more particularly from the Food and Agriculture Organization (FAO)—assistance in identifying the problems of the Peruvian fishing industry for which research was thought to be necessary, and in formulating a program to address those problems.

The result of that initial assistance from FAO was that, on April 21, 1960, a Plan of Operations was signed in Lima, under which, with financial support from the United Nations Special Fund, FAO was to give assistance to the Government of Peru in establishing an Institute for Research on Marine Resources. The project of assistance began operations on June 2, 1960, and continued for six years, during which slightly more than thirty-three man-years of expert time were spent in carrying out research and training local staff. The research was in fisheries biology, oceanography, fisheries technology, fisheries economics, and naval architecture. A second project of assistance operated for four-and-a-half years beginning July 17, 1967, and a third from July 1, 1972, for three years. In all, these three projects occupied more than sixty-seven man-years of foreign expert time, and a number of years of Peruvian staff time considerably greater than that of their visiting colleagues.

The objective of all these projects was the creation of a national capability to study the national fisheries in all its aspects—oceanographic, biological, economic, and technical—with a view to managing the exploitation of the resources to the best advantage of the nation. This objective was clearly formulated when, in 1964, the Council for Hydrobiological Research was merged with the Institute for Research on Marine Resources to form the Peruvian Marine Institute,

IMARPE, charged with planning, directing, executing, and coordinating research into fisheries. As will become evident in the sections below, the conduct of the Institute's work in the late sixties and subsequently (progressively in the hands of local staff until it was wholly so), is convincing proof of the success of that international assistance, as well as of the aptitude and dedication of the local staff.

At the beginning of the international collaboration, in 1960, the local and foreign staff were lodged in an old building, in Callao, which had previously been a kind of nightclub. Later, a four-story building was constructed to house the Institute, and in the 1970s this was extended. International assistance provided boats for research at sea, as well as for research on the Amazon and on Lake Titicaca; it also provided laboratory equipment, including a computer, and equipment for technological research.

It was only natural, throughout the course of the aid projects, and in subsequent years, that a large proportion of the research effort and of the use of facilities and services would be devoted to the anchoveta fishery. From its beginning, in the mid-1950s, the fishery grew rapidly, to take first place in the world as the one-species resource from which the greatest catch was taken. The anchoveta fleet reached its maximum size in 1970 with 1,499 vessels and a total hold capacity of 241,819 tons. Individual catches greater than 100 tons were commonplace. The primary question in the minds of both Government and industry was, "how large can, should, the catch be?" And this was taken to mean, "what is the maximum sustainable yield (MSY) of this resource?"

The first step taken toward arriving at some answers to this question was to ascertain accurately just how much was being caught, and at what cost (in terms of fishing effort). An early achievement of the local and foreign team was the establishment of a system of statistical record, with accurate tally being made, by the processing plant staff, of the quantity discharged by each boat. With this system was coupled a program of sampling the landed catches, to provide data on the size, age, and sex composition of the population. This work called for a detailed statistical analysis, conducted in accordance with ideas as to the dynamics of fish population under exploitation. The Peruvian operation was carried forward on a wave of enthusiasm for studies of population dynamics, and in particular for the models, then in mode, leading to estimates of MSY.

As these systems were being put into operation, the fishery continued to grow. On a single day during one of these years, the weight of anchoveta landed at the processing plants was greater than the weight of all catches of all species landed in all other countries of the South Pacific in the whole of that year. Over the same span of time, the capacity of the processing plants had so increased as to be equivalent to the total of all the fisheries of the world. The question became, "how far can this go?" To draw upon outside advice, the Institute convened meetings of experts to whom the Institute's research workers presented their data, the results of their analyses, and their interpretation of the evidence so scrutinized and sifted.

While the statistical and catch-composition systems gathered reliable information on the fish delivered to the processing plants, they could not measure accurately the whole of the quantity of fish killed by the fishing operations. There was, in the first place, the matter of the fish crushed in the hold, or smashed by the pumps, and reduced to a kind of thin bloody soup to be discharged at sea, and therefore its quantity could not be measured. There was also the matter of fish discarded at sea when a boat having become fully loaded could not take aboard a certain quantity remaining in the net. Some people estimated that in 1970 the landed catch, twelve million tons, was only four-fifths of the total actually caught. In addition to these two problems, there remained the matter of relating the landed catch to the stock of fish in the sea. The statistics indicated what had been landed, but could not show the proportion that landed catch was of the natural population from which it had been taken. Nor could the system, of itself, guarantee that the composition of the landed catch truly represented the composition of the population in the sea. In an endeavor to address some of these problems, the Institute undertook a series of operations that, perhaps, has not been matched anywhere in the world. These were known as the "Eureka" operations.

For one of the Eureka operations, the Institute would organize a set of twenty to thirty boats to make a lightning survey—not a *blitzkrieg*, but a *blitzblick*. Each boat would be assigned a course, due west from a set point on the coast, on which it was to travel out 50 to 60 or even 120 miles, turn ninety degrees and travel five miles north, make another ninety degree turn to travel back to the coast, and then repeat the performance. In passing along this transect the echo sounders would operate, and a watch would be kept for schools of anchoveta; some sample fishing could take place; water samples would be taken, to be brought back to the laboratory for analysis; water temperature would be taken regularly; plankton samples would be taken for study in the laboratory; sea-surface conditions and

wind and other meteorological records would be kept. Before a Eureka operation, the skippers would meet for a briefing; one or perhaps two observers would be placed on each boat, to do the sampling and keep the records. The field operation would be completed in forty-eight hours and within twenty-four hours all samples and records would be delivered to the central laboratory where an analysis of samples and data would proceed well into the night. Within a week, a preliminary report would be made to the Minister on the state of the fish stock: its apparent level of abundance, its distribution, and its composition, as well as a preliminary indication of the catch that might be expected for the immediate future. A complete report of all the work would be dispatched to the Minister within a month. The Eureka operations were carried out to provide real-time information—that is, information needed at the time that decisions had to be taken. In this respect, the work differed from much of contemporary fisheries research, which could give fishermen only some account, and perhaps explanation, of what had happened in the preceding year.

Presented with the results of on-going work, and especially the reports of those Eureka operations that had been undertaken since their last meeting, the experts weighed the evidence and offered opinions concerning the state of the stocks and the fishing regime to be considered—when to fish, and how much to take. Their expectations, or calculations, of the MSY varied from six to twelve million tons, but by 1973 the available evidence indicated that the total biomass of anchoveta, in the sea, was only about four million tons, having fallen from unexploited levels in the order of eighteen to twenty-four million tons.

Although the members of the Navy's General Staff were fully aware of the need for detailed and systematic research into all aspects of each fishery, and undoubtedly knew about the record of bird mortalities and the appearance of El Niño, it is unlikely that they could have had any clear idea of the complexity of all that happened at the time of that event, nor of what went before it and what followed after. Indeed, in the 1950s and 1960s there was no clear idea of what was going on during an El Niño event, nor even recognition of it as one event in a whole series of events; a misunderstanding which persists today in many quarters.

There was no mistake, however, in the belief that investigation of fishery resources would have to traverse many and diverse fields—in oceanography and meteorology, as well as in technology and economics—and it was in this sense that IMARPE developed. From its very beginning, the Institute ac-

Anchovies. One day's catch was greater than the weight of all the catches of all species landed in all other countries of the South Pacific during that entire year. Photo: IMARPE

corded much attention to physical, chemical, and biological oceanography, and made ample use of electronic fish-finding equipment in its work with regard to the anchoveta. The story of the Peruvian fisheries since 1970 has fully justified this polymathic strategy.

ENSO: the El Niño and Southern Oscillation Cycle

In the simple early view, El Niño was seen to be a warm "current" which flowed capriciously from the north, predictable only as to its appearance at Christmas time; moreover, it was thought to be purely local, a calamity private to Peru. How, in detail, the anchoveta population was affected, so adversely, by this warm current was not known, but the sequelae were held to be plain for everyone to see. This view is now known to be at least very incomplete, as is signified by the name given to the event of which El Niño is a part—the El Niño/Southern Oscillation (ENSO) cycle. Development of thought about this cycle can be traced in the issues of the Tropical Ocean-Atmosphere Newsletter published by the University of Washington. Until 1980, the Newsletter carried articles with titles such as "Appearances in the Western Pacific of Phenomena Induced by El Niño in 1979–80"—showing, nevertheless, a broadening in the view of what was happening. Later articles such as "The Southern Oscillation over the Indian Ocean," and articles connecting events over the Atlantic with the ENSO cycle, clearly indicated the breadth and magnitude of the system or systems of which El Niño is part.

As early as 1975, IMARPE was host to a United States multidisciplinary research project entitled, Coastal Upwelling Ecosystem Analysis (CUEA). The coordinator of that project was accommodated in the IMARPE building, and the Institute gave some logistical and other support. The visiting team comprised meteorologists, physical oceanographers, chemical oceanographers, and specialists in many divisions of biological oceanography.

The CUEA project was set up with objectives very close to the primary objectives of IMARPE. It was designed to give an account of the physical processes by which nutrient-rich water is brought to the surface from a great depth, of the processes by which the nutrients are spread at the surface and taken up by microscopic plant-life (phytoplankton), which is fed upon by small animals (zooplankton), and whereby both plants and animals are fed upon by fishes. The

Above: Coast of Hood Island
in the Galapagos chain. Photo
© 1991 Carl Roessler

Below: The research vessel
Humboldt, participating in the
Eureka operation. Photo:
IMARPE

primary physical process of upwelling is a result of wind and current forces operating in relation to particular physiographic features. As might be supposed, the strength of upwelling, in other words, the volume of water brought to the surface at a particular place (such as off Pisco on the southern Peruvian coast), varies at least seasonally, as winds vary from season to season and vary from year to year. To the variations in the physical systems are added variations in the cycles of living material and the behavior of fishes. Sometimes there is a matching of enrichment of surface waters with phytoplankton bloom, of that with zooplankton growth, and of the totality with the spawning of anchoveta and other fishes; sometimes the match is poor, with major effect on the fish populations. Obviously, an analysis of these systems could contribute importantly to an understanding of the anchoveta stocks, and if the oceanic events could be predicted, the anchoveta fishery could be managed with greater confidence, and with improved effectiveness.

An International Role

The CUEA project was operating at a time when the Peruvian anchoveta fishery was virtually at a halt, and the IMARPE staff felt themselves almost helpless against an El Niño event. It seemed obvious that there must be connections between the processes involved in upwelling, the effects produced by an El Niño event, and the changes in the stocks of anchoveta. It was clear that this entire great complex, which involved all four coastal states of the west coast of South America, could be analyzed and its behavior become predictable only by way of extensive international research. For these reasons IMARPE took the initiative to promote, under the auspices of the South Pacific Permanent Commission (SPPC), a collaboration among the four coastal states, and between them and other countries, in original research concerning these systems, as well as the establishment and operation of a cooperative effort to monitor the oceanic phenomena. In 1974, the Institute took a very active part in the Workshop on the Phenomenon known as El Niño, convened in Guayaquil by the International Oceanographic Commission (IOC) of UNESCO, and FAO. Various recommendations had been made at that workshop and in execution of them

a series of programs under the title "ERFEN" (*Estudio Regional del Fenomeno "El Niño"*) had been put into operation and discussed by the SPPC countries at meetings in IMARPE in 1975 and 1976. That workshop recommended support for the ERFEN project and made proposals for a coordination of international projects with the activities of ERFEN. The project ERFEN is now supervised by the South Pacific Permanent Commission, which regularly issues a Climatic Analysis Bulletin to which contributions are made by sixteen institutions of the four SPPC countries: Colombia, Chile, Ecuador, and Peru.

The SPPC effort is to be seen in the context of an extensive set of international programs that are, themselves, an orchestration of national programs. It is in the national programs, of course, that the substantive work of observation and measurement, analysis and interpretation, is carried out, but the coordination effected by the international programs, the intercalibration of instrument and method promoted by them, and the exchange of information that they further, are indispensable features of modern oceanography. An important segment of the international program, one of special relevance to the SPPC work, is the International Tropical Ocean Global Atmosphere Program, known under the acronym TOGA, news of which is disseminated in the Tropical Ocean-Atmosphere Newsletter (referred to earlier). A recent issue of that Newsletter had an article reporting that "the TOGA observing system" had been "successful in providing meteorological and oceanographic data . . . |which| have helped scientists to establish beyond all reasonable doubt that there is a predictability inherent in the ENSO cycle (though how much is not known)." It goes on to predict that, "in 1995 the capability will have been developed to make skillful predictions of large-scale tropical variability several months in advance," and to point out how important for the realization of that objective will be the national programs of *in situ* observation.

While that work is going on—and assuredly the collaborating institutions of the SPPC countries will be making their contributions—there will be matching research programs to better understand the relationships between the meteorological/oceanographic systems and the biological systems. In this work, IMARPE can be expected to play a major role.

11.

The Institute of Marine Sciences and Limnology, Mexico City

Agustin Ayala-Castañares

Adult Gray Whale in Magdalena Bay, Baja. This endangered species has enjoyed a remarkable recovery under the Mexican conservation program. Photo: © Eda Rogers/Marine Mammal Images

Introduction

This chapter is dedicated to the Institute of Marine Sciences and Limnology (*Instituto de Ciencias del Mar y Limnologia*, ICMyL) of the National Autonomous University of Mexico. This is a vigorous institution, created in 1973 to advance scientific research and graduate teaching. The most valuable resource of the Institute is represented by its academic personnel that were assembled after many years of uninterrupted efforts to train Mexican scientists in Mexico, and in other countries all over the world. As a consequence of this systematic training effort, significant research is now being carried out in Mexico and the capacity for indigenous training in the marine sciences has been well established. Research and training are supported by excellent facilities, including three coastal stations and two oceanographic vessels.

Mexico has large oceanic areas under its jurisdiction and an enormous potential for their development. The petroleum industry, supported by vast amounts of oil drawn from the continental shelf, is one of the pillars of the Mexican economy, and it is considered one of the largest in the world. Fishery production also plays an important role in the economy. The Mexican harbors are an essential part of the marine transport system and require strong support with a solid technical basis. There are also strategic submarine mineral deposits in the Mexican Pacific. Moreover, the coastal areas are well recognized for their extension, variety, resources, and beauty.

The country's scientific and technological tradition is limited; during the past two decades, however, it has vastly increased its capabilities. Since 1970, when the National Science and Technology Council (*Consejo Nacional de Ciencia y Tecnologia*, CONACyT) was established, the government has become much more aware of the importance of science and technology.

Efforts have been concentrated on building a scientific infrastructure and training highly qualified human resources. This was attained through long-term scholarship programs enabling young scientists to obtain their doctoral degrees in various institutions, both national and abroad, and by providing strong support to Mexican universities and centers of higher education. After 1970, through the joint efforts of UNAM, CONACyT, and other institutions, units specializing in

marine sciences were created, such as the ICMyL of UNAM, the Center of Scientific Research and Higher Education at Ensenada (*Centro de Investigacion Cientifica y Educacion Superior de Ensenada*, CIVCESE), the Center for Biological Research of South Baja California (*Centre de Investigaciones Biologicas de Baja California Sur*, CIB), and the Merida Unit of the Center for Research and Advanced Studies (*Unidad Merida del Centro de Investigacion y Estudios Avanzados*, CINVESTAV).

In addition, at great sacrifice, substantial investments have been made in facilities, vessels, and equipment. As a consequence, there is now a young, small, but vigorous Mexican marine scientific community properly prepared to contribute to the understanding and preservation of the ocean, to making good use of the Mexican marine resources, and to participation in international programs in the major marine sciences.

The National Autonomous University of Mexico is the most important national institution of higher learning, and it plays a critical role in the life of the Mexican nation. It has been the major source for the recruitment of the professionals required for the technological and administrative servicing of the country; it has also been able to allocate significant funds for research applied to national problems, and it has been responsible for extending cultural benefits.

Research at UNAM started more than a century ago, but was formally established in 1929 when the university's autonomous legal status was approved. At that time, three previously existing research institutions became part of the university. Currently, twenty-one research institutes and centers are devoted to the sciences, and many others to the humanities. UNAM plays an important role in the national system of science and technology, largely due to the significant funding it receives from the federal government.

Until recent years, university scientists devoted themselves primarily to fundamental research, but there is now a tendency to venture into applied research. Traditionally UNAM's research has been specialized—in biology, chemistry, geology, and physics—but during the last twenty years a trend toward establishing interdisciplinary units, such as the Institutes of Applied Mathematics, Materials Research, and Marine Sciences and Limnology, has developed.

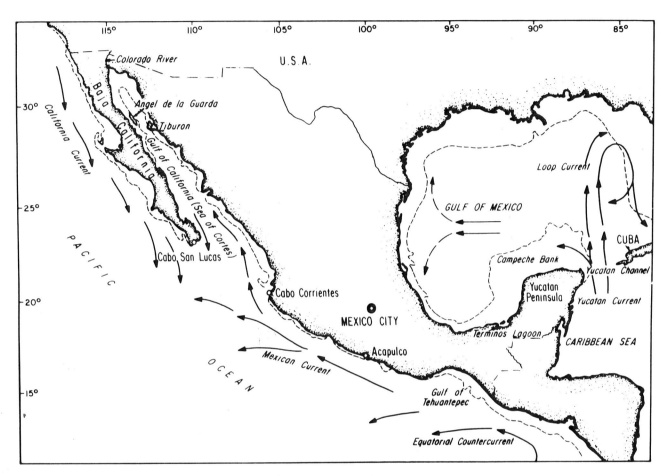

The Mexican Seas. Courtesy:
UNAM

The Mexican Seas: Legal and Political Aspects

Officially named the United States of Mexico (*Estados Unidos Mexicanos*), the country is also known as Republic of Mexico (*República Mexicana*), or, simply, Mexico. It is constituted as a "Representative Democratic and Federal Republic of Free and Sovereign States, Joined in a Federation." There are thirty-one states and a federal district within which Mexico City, the capital, is located.

The surface of the country is 1,958,201 square kilometers, including islands. Added to its territory are the Territorial Sea (231,813 square kilometers), and the Mexican Exclusive Economic Zone (2,715,012 square kilometers), delimited by a boundary of 188 miles offshore.

Mexico is delimited to the north by the United States of America, along a border of 3,326 kilometers; to the east, by the Gulf of Mexico and the Caribbean Sea, along a coast line of 2,905 kilometers; and to the west and southwest, by the Pacific Ocean (shore line: 7,338 kilometers).

According to the 1980 population census, the population of Mexico consisted of 66,846,833 inhabitants, and it was expected to exceed seventy-seven million by the middle of 1987. In the last decades, there has been large-scale migration from the country to the cities, especially to Monterrey, Guadalajara, and Mexico City, the latter being the most significant. The population of the Mexico City metropolitan area now exceeds seventeen million. Actually, the urban population represents two-thirds of the total. The official language is Spanish and it is spoken throughout almost all of the country, except for some small Indian communities. The major religion is Roman Catholic.

The international policy of Mexico is well known and, in accordance with the country's tradition, it is based systematically on certain fundamental principles: respecting non-intervention, and the sovereignty and self-determination of all States. For this reason, the Mexican Government strongly supports the United Nations system and multilateralism as a desirable mechanism of relations between countries.

The Mexican Government is aware of the political, economic, and social importance of the Mexican Seas. It participated for many years in the negotiations of the Third United Nations Conference on the Law of the Sea. Mexico was one of the countries that signed as soon as the Convention was opened for signature (December 1982), and in 1983 it ratified the Convention.

In 1976 the Mexican Constitution was amended to establish the Mexican Exclusive Economic Zone (EEZ). With this decision the area under Mexican jurisdiction was increased by 2,892,000 square kilometers; thus the marine portion is larger than the terrestrial.

In order to adjust Mexican municipal law to the letter and spirit of the United Nations Convention on the Law of the Sea, in 1986 the Mexican Congress promulgated the Federal Law of the Sea (*Ley Federal del Mar*): a body of laws almost as complex as the U.N. Convention itself, and, in many ways, a model for other developing countries.

Oceanographic Characteristics

The seas and oceanic areas adjacent to Mexico have been the sites of oceanographic and fisheries research during the last fifty years or so. This work was carried out mainly by foreign expeditions. However, in the last decade, contributions have come increasingly from Mexican researchers, particularly since the acquisition of the two research vessels of UNAM, commissioned at the beginning of the eighties.

Mexico is one of the most diversified geographical areas of the world, with deserts, tropical forests, snowcapped volcanoes, mountain ranges, and extensive lowlands. Rivaling the land area, the adjacent seas present a large variety of oceanographic conditions, with deep basins and trenches, as well as rugged and abrupt bottom relief.

Basically, the Mexican Seas consist of two parts: the Pacific and the Atlantic, which in turn may conveniently be divided into regional or oceanographic provinces. While such divisions depend, obviously, on the particular political or oceanographic aspect to be dealt with, it has been customary to adopt a general scheme consisting of three regions on either coast of the country, and the coastal lagoons.

It should be noted that, although the two parts of the Mexican Seas are situated within the same latitudes, both belonging to the warm water zone of the World Ocean, they differ from each other in many respects. This difference derives to a large extent from the fact that the waters on the Atlantic side are the end product of a surface circulation that originates in the upwelling areas off the African west coast, while the Mexican Pacific represents the starting point of a similar system that extends to the west across the tropical zone of the Pacific Ocean. Thus, the Atlantic side has a thick surface layer of warm and nutrient-depleted waters, in contrast to the thin, relatively cool, and nutrient-rich surface layer of the Pacific side. These conditions are strongly reflected in the living resources on either side of the country—the high organic productivity of the Pacific waters contrasting with the moderate production on the Atlantic side.

The three regions on the Pacific side are the Baja California Pacific region, the Gulf of California, and the Mexican Eastern Tropical Pacific Region. On the Atlantic side is the Southwest Gulf of Mexico, the Campeche Bank, and the Mexican Caribbean. On all three sides, there are coastal lagoons.

Each one of these regions has its own peculiarities, its problems, and its priorities, which may be very different in the coastal zone and in the open ocean.

The Baja California-Pacific

Situated along the west coast of the Baja California Peninsula, this region is dominated by the south-flowing California Current, as well as by upwelling of variable strength that takes place off the coast. Due to the upwelling and other favorable conditions, this region ranks among the most productive of the North Pacific, with rich marine life at all levels.

The annual presence of the migrating California gray whale (Eschrichtius robustus) in the coastal lagoons of Baja California is a notable feature; this species was once under risk of extinction. The mighty animals arrive in the lagoons during the winter months. They calve, nurse their calves, breed, and move on. The lagoons have been declared a sanctuary, and even tourism is severely restricted to ensure the tranquillity of the animals during this crucial time in their life cycle. As part of their protection and conservation policy, the Mexican National Gray Whale Program of the Secretariat of Fisheries has made systematic efforts in recent years to protect the environment as well as determine the abundance, distribution, and habits of this endangered species. These efforts have been quite successful, and have gained worldwide recognition.

Gulf of California

Also named the Sea of Cortes, this region may, in many respects, be considered an extension of the Tropical Pacific, and its waters originate mainly at lower latitudes. However, due to the local climatic conditions, the upper layer of the Gulf is subject to dramatic changes of temperature as well as salinity, particularly in its innermost portion. Since evaporation strongly exceeds precipitation and runoff from the surrounding land, salinity is increased to values about 35.5 per thousand, the highest in the North Pacific Ocean. The Gulf of California is indeed the only evaporation basin of the Pacific and, in this respect, it has been compared to the Red Sea. There is, however, a marked difference, caused by the fact that, while the Red Sea communicates with the open ocean through a narrow strait, the Gulf of California has a wide and deep opening to the adjacent oceanic areas.

The high salinity of this oceanic region has encouraged the establishment of a sea salt industry. The technology to extract salts from seawater is ancient. It extends back thousands of years and involves the oldest known use of solar energy. The seawater trapped at high tide in small basins, pools, holes, and cavities was simply allowed to evaporate, leaving behind salt crystals, either partly broken down into a white grainy mass or in sheet form, overlying a residue of seawater. All people had to do was collect the results of this natural process.

One of the largest sea salt mines in the world is at Guerrero Negro in Baja California, on the Gulf side. Covering over 49,000 acres, this solar evaporation system contains 200 kilometers of dikes, 40 kilometers of canals, and 45 kilometers of main haul roads. The endless expanse of salt—salt as far as the eye can see, salt heaped into mountains, salt everywhere—looks like vast fields of snow. The salt crystals crunch underfoot just like ice crystals. Mexico exports six million tons of sea salt annually to Japan.

The seasonal thermal range of the surface layer of the Gulf of California is also among the highest, with temperatures up to 30°C during the calm summer months, followed by a cooling down to 14°C under the prevailing northwest winds in winter. The tides in the inner Gulf are also intense, with a spring range of more than eight meters. Strong tidal currents, as well as mixing, may be observed in the inner Gulf, particularly near the Angel de la Guarda and Tiburon islands.

Through the vertical mixing and upwelling, nutrients are brought to the surface layer of the Gulf, stimulating primary production, a fact that is reflected in its abundant and highly diversified marine life—of which shrimp is presently the most valuable resource.

As has recently been discovered, the basins of the Gulf of California and the areas to the south of its entrance are studded with hydrothermal vents whose thermal and chemical effects on the surrounding water mass have not, as yet, been thoroughly investigated.

Joint research efforts have been carried out in

A hydrothermal zone in the Gulf of California. As recently discovered, the basins of the Gulf of California and the areas to the south of its entrance are studded with hydrothermal vents whose thermal and chemical effects on the surrounding water mass, and economic development potential, have yet to be explored. Photo: UNAM

this area by scientists of several institutions, including ICMyL, aboard the submarine *Alvin* of the Woods Hole Oceanographic Institution.

Mexican Tropical Pacific

Forming the northern portion of the Panamic Region, which extends from the equatorial zone along Central America and Mexico, this region is occupied by waters of tropical origin and its surface layer is characterized by fluctuating hydrological and dynamic conditions. The waters are moved toward the northwest by the Mexican Current and gradually feed into the North Equatorial Current of the Pacific. Upwelling is intense, particularly in areas like the Gulf of Tehuantepec where strong winds blowing off the shore move the surface layer away from the coast.

In this region a sub-surface layer of very low oxygen content extends along the coast with a maximum thickness and intensity in the area of Acapulco. Under strong upwelling, this hypoxic layer may ascend to some twenty meters below the sea surface. Consequently, in spite of abundant nutrients and high primary productivity, the benthic and neritic fauna is relatively poor over the narrow continental shelf of this region.

The Southwest Gulf of Mexico

This region extends along the east coast of Mexico from Rio Bravo in the north to Terminos Lagoon in the southeast, and out to the central area of the Gulf of Mexico. Near the coast the waters are influenced by the runoff from land, particularly in the southern portion where the river discharges may form wide, however thin, brackish plumes with sharp edges on top of the blue waters of the high Gulf. Under the influence of the prevailing easterly winds the waters are carried toward the coast to form a coastal flow of fluctuating strength and direction. Wide eddies, hatched by the Loop Current in the eastern portion of the Gulf, are often observed moving westward toward the coast of Mexico. Such eddies are believed to play an important role in the renewal of the waters that occupy the upper layers of the Gulf.

The bottom sediments are mostly of terrigenous origin. Near the shore, vertical mixing and the river discharges sustain a relatively high primary and sec-

ondary productivity. Here, as in the Gulf of California, the shrimp is the most valuable living resource. Petroleum basins and fields are found in the west-central as well as the southern-eastern portions of this region. It is in this region that the ill-fated blow-out and fire of the exploratory oil well *Ixtoc I* occurred in the late seventies—ICMyL made important contributions towards the interdisciplinary study of its ecological impact.

Campeche Bank

This province covers the wide and shallow shelf that extends some 150 to 250 kilometers toward the west and north from the coast of the Yucatan Peninsula. The average depth is around fifty meters, and there are several reefs and shoals spread over the shelf. The bottom is formed by calcareous rocks and sediments. At the eastern edge of the Campeche Bank, near the Yucatan Channel, there is an upwelling zone where deeper water ascends to the surface layer and, when flowing westward over the bank, provides nutrients that make this area one of the most productive parts of the Gulf of Mexico.

The most valuable living resources in this region are demersal fish, such as grouper (*Epinephelus*) and snapper (*Lutianus*), as well as spiny lobster (*Panulirus argus*) and octopus (*Octopus vulgaris*). The oil fields, recently discovered at the southwestern portion of this province, are now in full production.

The Mexican Caribbean

This region covers the Yucatan Channel and the westernmost part of the Caribbean Sea. The continental shelf is very narrow and the bottom is calcareous; the coastline is fringed with coral reefs that thrive in the clear and fast-flowing waters of the Yucatan Current. Due to their Atlantic and Caribbean origin, the surface waters of this province, depleted of nutrients, are of low primary and secondary production, albeit rich in variety and thus of high transparency which makes this area an ideal place for diving and visual observations of marine life in all its splendor. The most valuable living resources of this region are spiny lobster and pelagic migratory fish, such as tuna. However, its most appreciated asset is the natural condition for recreation and tourism.

Coastal Lagoons

Besides these sea and oceanic regions, extending from the coast to the limits of the Exclusive Economic Zone, there are over 100 lagoon systems that cover some 1.6 million hectares along 3,000 kilometers of the Mexican coastline, the total length of which is more than 10,000 kilometers of very variable origin, nature, and form. The coastal lagoons represent an important economic resource, direct as well as indirect, due to their fundamental role in coastal-zone development and the nearshore ecology.

The lagoons may have a temporary or permanent communication with the open sea, from which they are separated by barriers of diverse geological origin. The water regime of these lagoons is determined by the interaction between the runoff from land, on one side, and the penetration of seawater by tidal pumping and density currents, on the other. Since these factors vary from one area to the other along both coasts of the country, the coastal lagoons offer a wide diversity in characteristics. Seasonal variations are also strong, sometimes resulting in supersaline waters and closed inlets during the dry season, followed by brackish to fresh waters and wide open inlets in the rainy season.

The coastal lagoons are important nursing grounds for many species that live in the open sea and, therefore, are part of an ecosystem that extends far out into the adjacent marine areas. For this reason, any interference with the regime of the drainage area or the inlets of the lagoons may have serious consequences on both sides of the barriers that separate the lagoons from the sea. That is why a good deal of Mexican marine research efforts has been concentrated in this area of oceanography.

The Institute of Marine Sciences and Limnology ... Objectives and Functions

The objectives and functions of the Institute are determined by legislation and by university requirements. They are: to contribute to the knowledge of the oceans and continental waters, and their resources in general; to contribute to the scientific and technological development of the country, through the development of the marine sciences and limnology; to contribute to the knowledge of the Mexican seas and continental waters, and their resources; to participate and cooperate in the study and solution of problems of national significance, within its area of competence, in accordance with the organic law of the University of Mexico; to participate in the training of the scientific investigators, university professors, and highly qualified technicians required by the country; to establish the necessary relations with other academic units within the University of Mexico, as well as with other related institutions, through cooperation and advice; to stimulate the establishment of Institutions and Stations in Marine Sciences and Limnology in different zones of Mexico; to provide scientific and technical advice, both inside and outside the university; to form, maintain, and increase the scientific collections from the Mexican seas and continental waters; to contribute to the diffusion and diversification of the marine sciences and limnology.

... Historical Background

The Institute's origin may be traced back to research programs initiated in 1939 in the Institute of Biology under the leadership of the eminent Spanish scientist, Dr. Enrique Rioja Lo Bianco, who created the Section of Hydrobiology, conducted research, and prepared students. Those activities were extended in the 1950s to the Institutes of Geology and Geophysics.

These activities were carried out devotedly and with unflagging interest, but the available human and material resources were small, and the university lacked a clear policy with regard to oceanography.

In 1964, the Institute of Geology initiated a program of cooperation with UNESCO that included, as main component, the provision of long-term experts to assist the different institutes of UNAM in the various aspects of research, and preparation of human resources for research. That program started on a small scale and was increased gradually at UNAM. Later, it was moved to the Institute of Biology. It was the predecessor of the so-called National Plan of Science and Technology for the Creation of the National Infrastructure in Marine Sciences and Technology that was carried out between 1971 and 1974, under the auspices of the Government of Mexico (CONACyT) and the United Nations Development Program (UNDP), with UNESCO as the Executing Agency in the U.N. system. The ICMyL was the main Mexican institution involved.

The Department of Marine Sciences and Limnology of the Institute of Biology was created in 1967. Based on the Section of Hydrobiology, it had, nevertheless, an interdisciplinary character from the beginning and incorporated a group of scientists in marine geology who moved from the Institute of Geology to the Institute of Biology.

Aerial view of an oil well
blow-out in the Gulf of
Mexico. The Institute of
Marine Sciences and
Limnology was a major
contributor to the
interdisciplinary study of this
blow-out's ecological impact.
Photo: UNAM

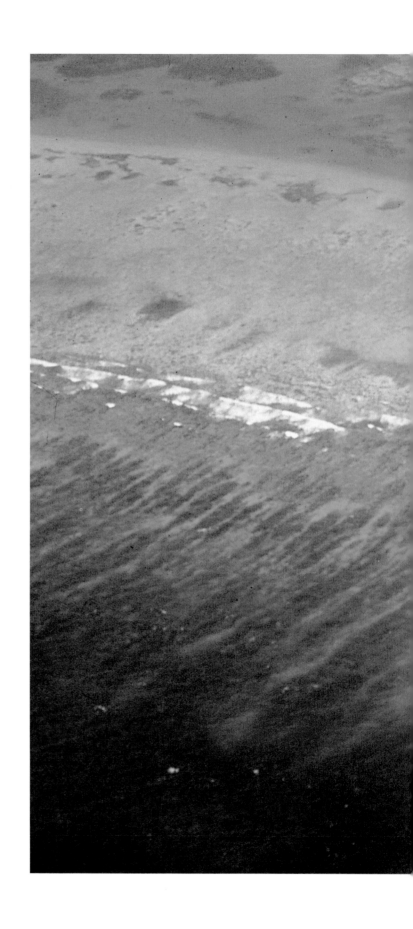

An overview of the coral reef
barrier zone. Mexico's
extensive fringed coral reef
areas are the second largest
in the world, after the Great
Barrier Reef of Australia.
Photo: UNAM

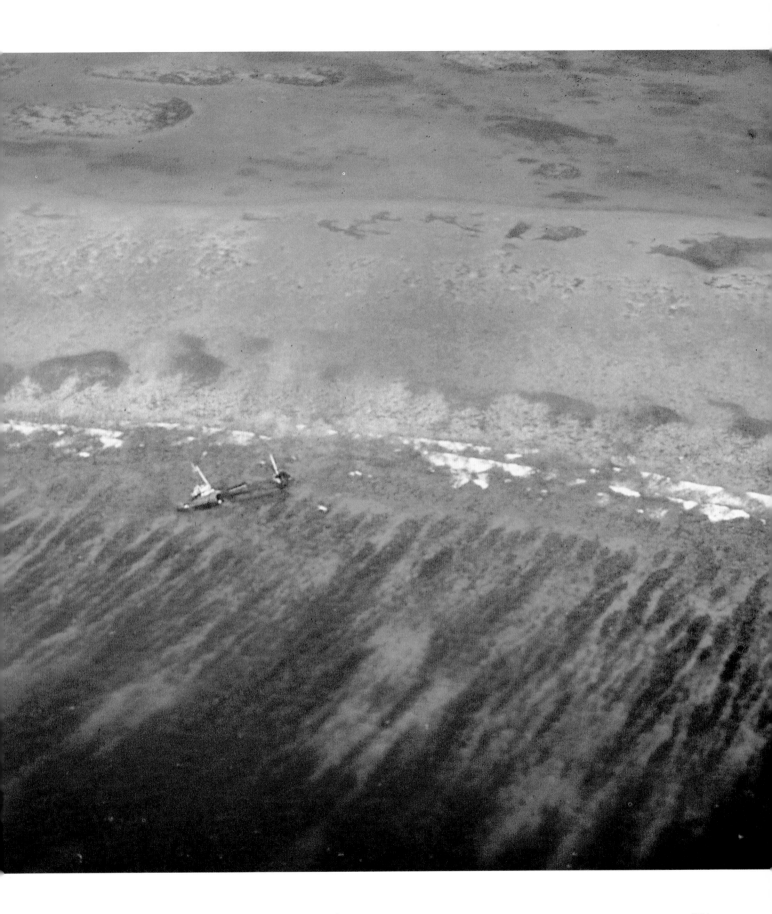

The new Department was transformed into a Center for Marine Sciences and Limnology in 1973. Its first and only Director was Dr. Alfredo Laguarda. The Center, in turn, was elevated to the status of an Institute in 1981 by the University Council. The first Director of the new Institute was the author of this chapter (1981–87). He was succeeded by Dr. Jorge Carranza.

... Research Areas

Maintaining its interdisciplinary character, the Institute covered physical oceanography (ocean processes and climate, coastal oceanography, oceanic modeling), as well as geological oceanography (including marine geology, sedimentology and diagenesis, paleo-oceanography, micro-paleoontology, foraminiferal and ostracod ecology), chemical oceanography (including marine chemistry, marine physical chemistry, marine pollution, and marine microbiology), biological oceanography and fisheries science (including echinoderm ecology, genetics of marine organisms, marine protozoology, phytoplankton and primary productivity, zooplankton, ichthyology and estuarine ecology, benthic ecology, marine pharmacology and fisheries biology), and, finally, limnology (that is, the study of freshwater resources and environment).

During the past ten years, all research areas have been strengthened, projects have been diversified, focusing, increasingly, on the solution of problems of national interest. The publications of the Institute's scientists are increasingly appreciated by the international scientific community. During the most recent years, a number of interdisciplinary projects have been carried out within a global or regional framework. Some of these projects have been financed by CONACyT.

... Academic Policy and Human Resources

There is now a well-defined academic policy. The primary goal is the improvement of the personnel. To reach a level of excellence, new researchers are engaged only if they hold a Ph.D., while the level of the existing personnel is being continuously upgraded. At present, the staff includes fifty-seven researchers, of whom thirty-six hold doctorates, nineteen master's degrees, and one a bachelor degree. There are also forty-eight technicians.

The Institute is concerned with the formation of human resources, both at national and regional levels. Ninety students, from different schools of the university and from other institutions, work on their theses in the Institute under the guidance of its scientists. The Institute's academic personnel teaches courses at different levels, from bachelor's to doctor's.

All the human resources of the Institute—including those assigned to its marine stations and the two research vessels—are used in the teaching program. The program is simultaneously taught in Mexico City and in Mazatlan. The annual student population entering the program ranges from eight to twenty-four students, with an average of ten; these individuals are very carefully selected by the Academic Council from among numerous applicants, both local and foreign. All applicants take a preparatory course and an examination on which final selection is based.

The program requires direct student participation in research activities, under the tutorial supervision of advisory committees consisting of three researchers at the master's level and five researchers at the doctorate level. Students receive a solid theoretical and practical formation in marine sciences, and they are prepared for original research, high-level teaching, and specialized professional work. A significant number of credits is given for participation in research projects, cruises, and field work.

The number of graduates is increasing each year. Over a period of eleven years, seventy master's degrees and twenty doctorates in marine sciences have been granted. Standards are high, in an international context. A number of graduates come from other countries—El Salvador, Costa Rica, Panama, Colombia, Ecuador, Peru, Chile, Brazil, Venezuela, and Argentina.

... National and International Cooperation

The Institute has an intensive program of cooperation, both at national and international levels.

Nationally, it has connections with higher education and research institutions in different parts of the country and it cooperates with various governmental organizations, such as CONACyT, PEMEX, various Secretariats (including Fisheries), and State Governments (such as Guerrero and Michoacan).

Internationally, there are bilateral and multilateral activities. Bilateral projects are carried out in cooperation with Australia, Costa Rica, Cuba, Germany, France, Great Britain, the United States, and the U.S.S.R. Multilateral projects include UNESCO and its intergovernmental Oceanographic Commission, as well as the Organization of American States (OAS).

The Institute's scientists have been involved in extensive scientific experiments, under bilateral and multilateral programs, such as the Joint Oceanographic Investigation of the Deep Sea (JOIDES) which brought the *Glomar Challenger* to Mexican waters; the study of hydrothermal vents, bringing *Alvin* twice to the Gulf of California; they have participated in the planning phase of the World Ocean Climate Experiment (WOCE), and in the Group of Experts on Ocean Sciences and Non-Living Resources (OSNLR), as well as Living Resources (SLR), particularly in the Caribbean region.

In 1988, the Joint Oceanographic Assembly was held in Mexico. Mexican scientists made major contributions, particularly in the Symposium "Oceanography in Mexico," and they provided a good demonstration of the capacity and potential of Mexican marine science.

. . . Facilities and Research Vessels

To enable the Institute to carry out its activities in the various coastal areas, facilities are highly decentralized. They include *Ciudad Universitaria* in Mexico City, *Mazatlan* Station, *El Carmen* Station, and *Puerto Morelos* Station, as well as the research vessels *El Puma* in Mazatlan and the *Justo Sierra* in Tuxpan.

Headquarters are in *Ciudad Universitaria*, that is, UNAM's main campus, in Mexico City. The administrative offices and main installation of the Institute are located there. On campus, the Institute has access to the general infrastructure of the university, including the possibility of using if needed, the laboratories and services of other institutes, centers, schools, libraries, etc., which may be required for the broad and complex interdisciplinary activities conducted by the Institute. It also has its own technical services, such as computing, electronic microscopy, and photography.

Mazatlan Station is located at Mazatlan, Sinaloa, in the northwestern part of the country, on the Pacific Ocean at the eastern side of the entrance to the Gulf of California, in the middle of a region of extensive coastal lagoons, with a strong interest in fisheries, harbor development, agriculture, tourism, and coastal area development. Laboratory space is available for geological as well as chemical oceanography. Marine pollution labs and facilities for the study of marine microbiology, of phytoplankton, zooplankton, benthic ecology, shrimp biology, ichthyology, and basic aquaculture (including an experimental aquarium room with running seawater), are housed in the three-building unit.

El Carmen Station is located at Ciudad Del Carmen, Campeche, in the southern part of the Gulf of Mexico, where the transition between the terrigenous province of the Southwestern Gulf of Mexico and the calcareous province of the Campeche Bank occurs, with access to the Terminos Lagoon (the largest coastal lagoon in the Gulf of Mexico) and the Gulf of Mexico itself. The three-building unit installation has four laboratories for multiple use, as well as accommodations for visiting scientists and graduate students. There is a small pier available for crafts going to the Terminos Lagoon. The *El Carmen* Station gives access to work in the coastal area and offshore of one of the most important regions of Mexico. It is the area with the largest oil production in the country, as well as being a vital center for fisheries, forest, commerce, and agriculture. The Terminos Lagoon area has been intensely studied by scientists of the Institute in collaboration with colleagues from other countries; it is one of the most attractive areas in the tropical coastal ecology of the world. Numerous publications by the staff of the Institute have been devoted to research carried out in this area.

The *Puerto Morelos* Station is located at Puerto Morelos, Quitana Roo, on the Mexican Caribbean, twenty miles south of Cancun in an area of fabulous touristic potential. The station, with excellent laboratory spaces and living facilities for visiting scientists and graduate students, has easy access to the coastal zone of the region, as well as to the extensive fringed coral reef areas (the second largest in the world, after the Great Barrier Reef of Australia), which provides unique conditions for studying those attractive and important environments.

In addition to a number of small craft, the Institute operates two medium-sized research vessels specially designed and built for multiple research and training activities.

These vessels are the R/V *El Puma*, with its own pier and operational unit, based in Mazatlan, for work in the Pacific Ocean that was initiated in 1981, and the R/V *Justo Sierra*, based in Tuxpan, for operation in the Gulf of Mexico and the Caribbean Sea, beginning in 1983. They are the property of UNAM, and operational costs for both vessels are shared between UNAM, CONACyT, and PEMEX through an inter-institutional, national cooperative agreement.

The *Mazatlan* Research Station of the Institute of Marine Sciences contains pollution laboratories and facilities for the study of marine microbiology, phytoplankton, zooplankton, benthic ecology, shrimp biology, ichthyology, and basic aquaculture, including an experimental aquarium room with running seawater. Photo: UNAM

The *El Carmen* Research
Station is a three-building
unit installation with
laboratories for multiple uses,
together with
accommodations for visiting
scientists and graduate
students. Photo: UNAM

The decision to acquire ships for UNAM was made after several years of careful consideration. It was clear that, before undertaking such a costly acquisition, several basic conditions had to be fulfilled. The most important among these was that there should be a demonstrable need for such research vessels. The second condition was that an adequate academic infrastructure and a supply of manpower to fully utilize such costly scientific tools should be readily available.

When considering the type of vessel to be designed and built, many aspects were taken into account. Among these, versatility and scientific work space were given a high priority, along with low crew requirements, adequate accommodations for scientists and students, and a reasonable cost of operation.

During the selection process, several offers of "donations" particularly of old foreign navy vessels, were studied. These were all rejected, since experience in other countries showed that adapting an old vessel in a satisfactory way and running it under given circumstances is more expensive, in the long run, than operating a new vessel designed to meet the particular needs of national marine research and the conditions under which that vessel will be used.

When UNAM decided to acquire a ship in 1979, intensive studies and consultations were undertaken to work out the best design and equipment, taking into consideration the operational needs, as well as the availability of spare parts and special maintenance services.

The vessels contain ample work space in the form of laboratories, mechanical and electronic workshops, a recording room, a lecture room, and the like. Given their advanced basic equipment and general laboratory, as well as deck capacities, these ships are well adapted to the multiple tasks of modern marine research.

. . . Scientific Journals

The results of the Institute's work are published in a number of national and international journals. The Institute also has its own organs of publication, the *Anales del Instituto de Ciencias del Mar y Limnologia* and the *Publicaciones Especiales del Instituto de Ciencias del Mar y Limnologia*. Another series was recently initiated, *Cuadernos Tecnicos del Instituto de Ciencias del Mar y Limnologia*.

The *Anales* is a periodical, arbitrated journal with international distribution, published annually since 1974. The *Publicaciones Especiales* is a non-periodical journal, also with international distribution. Thus far, nine issues have been published.

Conclusion

The development of science at UNAM and, in particular, at the Institute of Marine Sciences and Limnology, has been impressive and of great benefit to Mexico. This progress is the result of sustained long-term efforts involving several national institutions, as well as international organizations, and spanning five six-year administrative periods of the Mexican Government.

In recent years, unfortunately, the general support for science and technology in Mexico has had to be drastically reduced due to the economic crisis. Among the consequences of this difficult situation has been a substantial reduction in scientists' real salaries, in new equipment and funds for equipment maintenance and operations, library acquisitions, travel and scientific exchange, as well as new investments in scientific infrastructure.

This situation has sent a shock wave through the Mexican scientific community. Hopes have been dashed and the brain drain has intensified. A failure to reverse these trends soon may have serious consequences for the country's future development.

A major challenge for all of us is to maintain this Institute and to make a concentrated effort to strengthen its capacities and continue its scientific progress. Everyone recognizes an historical responsibility to find ways and means for providing adequate funds to consolidate the Mexican scientific system and to use science and technology to satisfy the evident demands of the country in accordance with current and future needs. Only time will tell whether or not we have been able to meet the challenge successfully.

The research vessel El Puma, based in Mazatlan, near the entrance to the Gulf of California. The main considerations on this vessel are versatility and scientific work space, together with low crew requirements, adequate accommodations for scientists and students, as well as a reasonable cost of operation.
Photo: ICMyL

12.

The Kenya Marine and Freshwater Research Institute, Mombasa

Eric Onyango Odada

The Kenya coastline. Kenya's Indian Ocean coast stretches for more than 600 kilometers just south of the equator. Photo: © 1987 Thomas D. Mangelsen

Introduction

The people of Kenya have had a long relationship with the ocean that washes its shores. Text fragments, the logs of ancient mariners, and the maps of flat-earth geographers enable us to trace it back to the times of the Greeks and Romans. Herodotus mentions a voyage around Africa that reportedly took place in the reign of King Necho of Egypt (609–595 B.C.). On Necho's orders, the Phoenicians, who were great seafarers, were to have sailed around Africa from the Red Sea to the Indian Ocean. Although the accuracy of this and other stories of similar voyages cannot be proven, there are records which show that Arab ships have for centuries sailed across the Indian Ocean from East Africa to Asia and back again.

The early travelers to the Far East were carried along by strong winds that blow across the Indian Ocean. They noticed that for four or five months the winds would blow steadily from one direction and then change, blowing for about the same period from the opposite direction. Many of these early travelers were Arabs, and the word "monsoon," which means "season," was used to describe these winds which were to play such an important role in the life of the continent. Travelers could plan to come from East Africa to India, Persia, and Arabia with the southwest monsoon and return later in the year with the northeast monsoon. Goods, such as spices, ivory—not to mention slaves, which were the main exports from East Africa—could be transported to the Far East and the products from that region could be brought back on the return voyage. Even today, dhows come to East Africa from Arabia to exchange goods. Thus, for a long time, the oceans around us have played an important role in the economic life of Kenya.

However, for centuries, the African people knew little about the oceans, and what they knew was mixed with myths and fables. Therefore, they had their own explanation for the monsoons. As reported by Warren, Africans believed that the seas and oceans are entrusted to the care of an angel. When he immerses the heel of his foot into the sea at the far end of China, the sea rises and flows westward. When he raises his foot from the sea, the water returns to its former place. This explains the surface currents of the North Indian Ocean, that change direction with the monsoons. As

the story goes, at times, the angel puts only the big toe of his right foot into the water—this probably referred to the cause of the semi-diurnal tide.

Stories of this kind show that the oceans have always captured the curiosity and fired the imagination of the African people. They have also been well aware that the oceans are as dangerous as they are bountiful. To utilize them properly, however, it was necessary to comprehend their real nature. In recent decades, African nations have begun to give priority to ocean development. They have started to invest in marine research and development programs aimed at the exploration and exploitation of marine resources such as food, minerals, energy, transportation, and recreation to improve the socioeconomic conditions of the African people.

The Sea Around Us

Kenya's Indian Ocean coast stretches for more than 600 kilometers just south of the Equator. Its silvery beaches glistening in the sun are shaded by whispering palms to form one of the world's truly great holiday playgrounds—deep-sea fishing, water skiing, sailing, surfing, and much more are available. Today, thousands of tourists from all over the world flock to this tropical paradise where a warm zephyr breeze blows gently all the time. Here, too, is a magical mixture of timeless cultures, traditions, and customs, reflecting two thousand years of history.

The parallel reef is close to the shore—in some places not much more than forty-five meters from the land, in an almost continuous escarpment sweep-up to the black headlands. There is not much water for swimming in the shallows, except at high tides. But it is possible to paddle out across the exposed coral, looking for shells and forms of minor marine life that have become stranded in the rock pools. When it is calm, a special attraction in the area is the opportunity to dive off the outer edge of the reef, and "goggle" or scuba among the pelagic fish—kingfish and kole-kole—sweeping past the sloping face and back into the blue haze of the open ocean. Unicorn fish glide away from the diver, large rockfish dart back into their coral holes, and the lobsters wave their long feelers. Big turtles come to feed on the lower slopes, but on the whole they are much shyer than the fish and soon paddle off when approached. All in all, it is one of the world's finest diving areas.

At sea, an observer might notice in the early evenings, when the sun is red above the dust of the inland bush, the low shafts of light that wash the horizon like a play of pink and amber colors in a theater, but with no framing proscenium and no other scenic intrusion: only vast, untrammeled sky. The effect is even more dramatic if there is a threat of rain in May or October, when there may be a rim of gilt around the heavy storm clouds. The low light also flares off the breakers across the reef that is unbroken for much of its length.

In places, the approach to the outer shelf is exposed when the tide is out, but elsewhere there are pools and channels for swimming. The beach is a gentle gradient, from forty-five to ninety meters wide. There may be scattered patches of seaweed about and a thin tidemark of debris, including sculptured driftwood. But generally the sand is clear, fine grained, with the texture of sea salt, and roughly the same color. When the beach is deserted and absolutely still, apart from the scurrying ghost crabs and a procession of sandpipers at the waterline, it conveys a sense of void—void of everything but the elements of air, water and earth. But then the lamps are switched on along a line of palms before the beachfront hotels. Nightfall occurs at about seven o'clock in the evening. In all respects, Kenya's coast is a superb holiday destination.

National Marine Parks

The Watamu and Malindi Marine Parks cover about 210 square kilometers of inshore waters and extend out to the old 4.8 kilometer territorial limit. Watamu sanctuary is shallow and uneventful, apart from a channel and line of coral gardens running parallel to the shore. Much of it is lumpy brain coral, not altogether favored by the shoals of reef-fish that prefer the protective branches of the madrepores, finger, and staghorn. But clown fish are well represented and so are sea anemones, which may form a complete, colorful tent over a head of brain coral. Occasionally a shapeless brown blob turns out to be an octopus. They are normally extremely shy creatures that flee at the slightest disturbance: but, like the other fish, the octopus in the coral gardens seem much tamer than elsewhere, and merely slip quietly away.

"Whale Island," located inside the southern boundary of Watamu Park, may be the chief attraction. This formidable, jagged hump of coral is partly overgrown with grass and low scrub, with the bulk of its "body" separated from the "tail" by a narrow sandy cove. At low tide, it is possible to walk around the base of the island, perhaps foraging for eels, but in any event viewing the minor marine life—sea urchins, rock crabs, and lizards.

Beyond the Watamu section, the reserve extends up the coastline to incorporate the Malindi Marine National Park. This comprises a series of reefs, roughly parallel to the shore. There are also a number of deeper-water channels and passages, a set of coral gardens and a sand island. The reef here is covered with various corals, including the madrepore species with their mauve or pink tips, and the curious saucerlike mushroom corals. Over and around more static creatures flows a bewildering variety of fish of every conceivable shape and color. At first, the observer retains only hazy impressions—a shifting blue-green veil of demoiselles, a splash of yellow butterfly fish, the brilliant blue and black streak of a cleaner wrasse—blue and yellow, red and green striped, spotty or blotchy.

Mombasa Town

Mombasa is not normally a specific destination for people on holiday. But it is always there, steamy, full of an atmosphere of its own and genuinely friendly, for a day's break from sunning by the ocean. In the old days, all arrivals, of course, were by ship—settlers, soldiers, and the first tourists, such as young Winston Churchill in 1907. All he had seen in his travels down from the Mediterranean were "the hot stones of Malta," the cinders of Aden, more scorched earth in northern Somaliland: and then Mombasa, "alluring, even delicious as she rises from the sea." The Churchillian mind "saluted," so he said, "with a feeling of grateful delight at these shores of vivid and exuberant green."

It is not quite like that these days, since the island's fifteen square kilometers of rising green is now almost entirely covered by the town. But the old town with its monument relics is still there: the Portuguese "Fort Jesus" brooding over the waterfront on its coral bluff. The great fortress is intact, with its massive walls, sand-ochre in color and blackened in places as though the powder burns of ancient battles had yet to be cleared off.

But it is quiet today, a museum reflecting its own

The old town, Mombasa.
Photo: KFMRI

violent and varied history. Inside, the story is prefaced by an exhibition of bits and pieces of pottery and other relics of the trade life of the coast from the ninth century to Vasco da Gama in 1498. Thereafter, the running narrative is detailed in old records, artifacts, and the interior structures of the fort.

The collection includes plans of the original construction in 1593 and of large-scale "indestructible" extensions that continued until 1631—as well as evidence of the first demolitions by hostile Arabs. Defunct cannon and other naval artillery are used to underline the various bombardments of Mombasa and the fort, from Ali Bey, the Pirate's, impertinent broadsides in 1585 to the last rounds fired by the British Navy three centuries later in 1875 to dislodge a rebellion against the ruling sultan of Zanzibar.

Mombasa has weathered all these storms, and with the new commerce and communications of the twentieth century it is the old port of Mombasa that has retained the dominant position. It is Kenya's premier port, catering to the needs of most eastern and central African countries. It is also the headquarters of the national oceanographic institutions. But most of all, it is a living embodiment of the relationship between the Kenyan people and the oceans around them.

The Kenya Marine and Freshwater Research Institute

Scientific research is a powerful tool for accelerating the pace of development, and no nation can afford to neglect it. In Kenya, research and experimental development are being utilized increasingly to serve the socioeconomic needs of our society.

The need to link research to national development has led the Government of Kenya to create a number of institutions under the Science and Technology Act of 1980. The KMFRI is one of the institutions established under that Act to undertake research in the field of biological, physical, chemical, and geological oceanography, as well as meteorology. The main objective of the latter is to correlate weather parameters with fish availability and fishing activities. In addition, the Institute conducts research on the socioeconomic impact of exploration and exploitation of aquatic resources. It is a good example of the importance Kenya attaches to a better understanding of the nature of the oceans and their resources, and to the need to base Government socioeconomic planning on scientific information. The research findings of the Institute are communicated to the Government through the National Council for Science and Technology (NCST), which advises the Government of Kenya on matters relating to science and technology for development.

Prior to the establishment of the KMFRI in 1980, marine and freshwater research was carried out under the auspices of the now defunct Research Organization of the East African Community (Kenya, Uganda, and Tanzania). After the collapse of the community in 1977, each country decided to proceed on its own.

Objectives and Programs

The research programs of the KMFRI are divided into two major groups, namely, the marine and the freshwater sectors.

The main objective of the marine sector, situated at Mombasa, is to collect and collate all available fisheries resources data. In addition, the Mombasa Laboratory carries out baseline studies of biomass and primary productivity of the coastal waters, distribution of nutrients in inshore waters, circulation in the inshore and adjacent offshore waters, coastal processes such as erosion and sedimentation, marine pollution, and mangrove ecosystems.

The freshwater sector, with two main laboratories and two substations, at Kisumu, Turkana, Sangoro, and Gogo, respectively, is engaged in biological studies of fisheries and environmental monitoring of inland lakes and rivers. In addition, these laboratories are conducting studies on aquaculture and limnology, as well as on pollution from industrial and agricultural wastes that find their ways through the rivers into the lakes, the main source of freshwater fish in the country. Kenya has over 10,000 sq kilometers of inland waters.

Human Resources

KMFRI has a total of 275 employees. Of these, 45 are researchers, 30 are technicians, and 200 are administrative and support personnel. Of the 45 researchers, Mombasa Laboratory has 26, Kisumu and sub-stations have 18, and one researcher is based at Turkana Laboratory, together with other scientists from abroad. Many of the members of the scientific community are young, and a significant number of them are engaged in postgraduate studies both at national universities and in other countries. Some are doing their doctorate work in oceanography abroad.

Small fisheries research craft.
Photo: KFMRI

The Kenya Marine and
Freshwater Research Institute.
It conducts research on the
living resources of the sea,
and coastal processes such as
erosion and sedimentation,
marine pollution, and
mangrove ecosystems. Photo:
KFMRI

Research Vessels

In addition to a considerable number of small sailing
craft, KMFRI operates two medium-size vessels specially designed for multipurpose research activities.
The *Maumba*, stationed at Mombasa Laboratory, is fitted with trawling and oceanographic winches and is
capable of deep-sea work. A similar vessel, the *Utafiti*, is
based at Kisumu Laboratory on Lake Victoria and is
specially designed for research in lakes.

Other Facilities

KMFRI's main laboratories at Mombasa, Kisumu, and
Turkana each have facilities for visiting scientists. The
Mombasa Laboratory has computing facilities with
KAYPRO I, OLIVETTI M21, and APPLE III microcompu-
ters fitted with printers. There is also a documentation
center at Mombasa, with facilities for retrieving infor-
mation from selected centers in Europe and the United
States. The center is linked to other marine research

institutions in neighboring countries and with the UNEP center for information in Nairobi. The center is used for the retrieval of scientific literature requested by the scientists through a selective computer system from the University of Liège, Belgium—about 30,000 pages of scientific literature connected with the ongoing research have been acquired.

International Cooperation

The achievements of KMFRI, although national in scope, would have been impossible without international support and cooperation, both from bilateral and multilateral sources.

Multilaterally, UNDP, UNESCO, and UNEP have given substantial support to the Institute, especially in the field of research training and the provision of equipment. A considerable number of research personnel, including this writer, have benefited from fellowships awarded by these agencies of the United Nations.

As for bilateral cooperation, a number of projects have been undertaken with help from NORAD (Norway), USAID (the United States), SIDA (Sweden), CIDA (Canada), the British Council, and Belgium. The latter donor is at present supporting an artemia production project along the coast of Kenya. The biological oceanography program, which is carried out by both Kenyan and Belgian scientists, also covers chemical studies of the inshore waters, mangrove ecosystems, and oyster culture in the creeks.

Fisheries Resources Development

This is the main area of KMFRI's specialization. It is, of course, a priority area for many marine research laboratories in developing countries. The importance of fish as a source of animal protein has been generally recognized, and many countries throughout the world are engaged in intensive programs to develop their fishing industries. In Kenya, fish and fish products are being developed to combat malnutrition. The main problem, however, has been the lack of the modern technology that the industry requires to function efficiently.

A survey of the marine fisheries resources of Kenya began in 1951, when the East African Marine Fisheries Research Organization (EAMFRO) was formed. Williams reported that about twenty-two percent of the total catches between 1951 and 1954 were *scomberomorus commerson* (seerfish). He also observed

that tuna was present throughout the year, but with a marked increase during the southeast monsoon and then very close to the shore.

Catch rates have been 8.87/100 hooks/per hour, and at the peak of the season during the northeast monsoon, catch rates have reached 9.2/100 hooks/per hour, with an average weight of forty-five kilograms per fish. The majority of the catches have been made below the thermocline (22–23c). These heavy concentrations during the northeast monsoon are associated with post-spawning feeding migration.

Fishing is mainly confined to the coastal waters up to fifty meters in depth. Only recently has it been extended to grounds up to 200 meters for deep-water lobster, prawns, and demersal fish.

Demersal fish now make up more than sixty percent of the total catch, followed by pelagic fish, some crustaceans, and a very small amount of mollusks.

Statistics are inadequate and out-of-date, but the general trend is clear: catches have been increasing moderately, but the resource is still largely unexploited. To realize its full potential, more research and more exploration is needed, fishing gear and methods must be improved, fish handling, processing, and marketing—as well as fish conservation and management—must be developed in order to increase production without the depletion of fisheries resources.

Fisheries Resources

Marine fisheries off the coast of Kenya fall into three categories. These are reef and inshore, continental shelf, and open ocean beyond the continental shelf.

Coral reefs and inshore waters around reefs are inhabited by many species suitable for commercial exploitation, but the nature of coral reefs makes the catching of large quantities of fish difficult. It is necessary to resort to techniques such as handline, traps and spear fishing, which are inefficient compared to the use of nets in trawling.

Fisheries from the continental shelf are almost inaccessible to Kenya fishermen. The continental shelf is extremely narrow, the edge is only about four kilometers from the shore, and the coral outcrops are hazardous in this region to any net used in trawling. In addition, hydrographical conditions work against local nearshore fishermen, since currents up to four knots are too strong for sailing craft, and motorized craft are too expensive for the local fishermen.

Open-ocean fisheries offer the greatest poten-

Oyster development project.
Oysters are grown on
mangrove poles. Photo: KFMRI

A mangrove forest. Mangrove swamps have been assigned a high priority among the topics under investigation by the Institute. Photo: KFMRI

tial, especially for tuna and other seasonally migrating fish, but the problems of marketing are compounded by the hot climate, the need for a system of distribution to major centers of population far from the sea, eating habits, and the possible effects of competition with well established freshwater fisheries on the lakes of the Rift Valley and Lake Victoria. The hydrography of the region may also present problems to open-sea fisheries and requires more study. All possible information on the manner in which the local currents and physical conditions of the water affect the breeding and recruitment of fish stocks must be investigated.

Other areas requiring investigation in relation to fisheries resources development are mangrove swamps and marine pollution. Mangrove swamps are the breeding grounds for many species of fish and crustaceans. Pollution, obviously, constitutes a severe threat to marine habitats and affects the distribution of organisms.

Mangrove Swamp Studies

Mangrove swamps are important both as a resource and as an ecosystem, but very little is known of the mangrove ecology of Kenya. Mangrove swamps have thus been given a high priority among the topics under investigation by the Institute. The mangrove ecosystem is one of the most productive in the world and plays an important role in the ecology of nearshore waters. Many exploitable organisms such as shrimps and crabs are associated with swamps. In addition, mangrove swamps often constitute the breeding grounds, or rearing grounds, for larvae of commercially important marine fish and crustaceans normally captured at some distance from the shore. Investigations conducted by KMFRI thus include the temporal and geographical variations in the distribution of larval, juvenile, and adult shrimps and crabs in and off the mangrove swamps. Such investigations involve seasonal plankton surveys in the field, and the culture of organisms in the laboratory to enable identification of the developing stages in later field programs. In addition, the in-and-out migration of larvae and adult local fish associated with the mangroves must be studied.

The mangrove trees themselves are cropped both for firewood and for timber in Kenya. The latter is used by the building industry and mangrove poles are exported by dhows from Lamu to the Arab countries, earning the country much needed foreign currency. In recent years, the rate and variety of human activities have increased to the point where a large proportion of the coastal mangrove resource is threatened with destruction. Unlike natural disturbances, such as damage from storms or tidal waves, human activities often induce changes in the intertidal environment so that subsequent recolonization and reestablishment of the original vegetation is impossible. In Kenya, water diversions for power and agricultural purposes are severely disturbing mangrove ecosystems by diminishing freshwater flows, increasing soil salinities, and interfering with nutrient supply. These human uses are often in direct conflict with the objectives of conservation of the mangrove ecosystem. They eliminate the protection provided by these systems to the coast and the inland coastal zone ecosystems; they eliminate the protection provided to the adjacent estuarine and nearshore marine ecosystem; they eliminate the habitat for many forms of wildlife, especially birds, that have important cultural, touristic, or scientific values; they eliminate the renewable timber, fuel, and food resources; and they eliminate the biological species variety of plants and animals, bound together over a long period of evolution, and still imperfectly known and understood.

Timeless, traditional interest in the mangrove swamps has caused them to be incorporated into the legends of many cultures of our coastal communities. KMFRI's objective, therefore, is to study first the uses made of mangrove ecosystems, including the attitude and behavior of the coastal people who have a long historic and biological dependence on the mangrove swamps and forests, and to try to resolve the conflicts arising from these various uses. Another objective of KMFRI is to investigate the properties of the mangrove systems that enhance their value for marine fisheries production, coastal protection, forest production, and biological conservation.

Marine Pollution Studies

As a tourist attraction, the coast of Kenya is renowned worldwide, and the internationally famous marine parks at Watamu, Malindi, and Shimoni are treasures that must be guarded through careful monitoring of factors such as freshwater runoffs, with their associated sedimentation and low salinity that pose a threat to growing coral. Marine pollution monitoring aims at identifying causes and effects, such as the selective demise of certain coral species or other shifts in the floral and faunal population. The main objective of such monitoring, therefore, is to ensure the maintenance of a high standard of tourist attraction and, at the same time, provide opportunities for academic research into the coral environment.

Marine pollution off the Kenya coast comes from various sources. The most obvious pollutant is oil spills. The introduction of oil into an enclosed marine habitat such as Mombasa Harbor can gravely affect the distribution of organisms. In addition to oil there are other effluents that are potentially dangerous to marine habitats and they are being monitored carefully by KMFRI. Untreated sewage, causing algae blooms that suffocate other marine life, is becoming a cause of concern. An increasing burden of poisons, such as heavy metals or organic pesticide residues like organochlorines, are carried into the nearshore waters, and the nature of our climate, with seasonal heavy rain, causes much topsoil from agricultural land to be washed into the Indian Ocean at certain times of the year. A direct route thereby exists for poisonous substances to be incorporated into marine sediments that build up in estuaries and adjacent areas. To identify and/or predict toxic effects on the marine ecosystem, monitoring programs have been initiated by KMFRI along the coast and offshore to study the gradients of such toxic compounds in estuarine and marine sediments.

Coastal Management

Environmental studies related to Mombasa Harbor development have recently become one of KMFRI's main specializations. Mombasa has a large natural harbor much of which is still underdeveloped and thus the natural environment has been preserved, to a large extent. It is an important entrance to Kenya and the landlocked countries in Central Africa. The development of the harbor, therefore, has been accorded a very high priority in present and future planning.

KMFRI's main concern is to ensure that the authorities planning the harbor development give an equally high priority to protecting the fine coral reefs and sand beaches. To protect this environment, developments should be sanctioned only if comprehensive environmental impact assessment studies have been completed.

A study commissioned by the Kenya Port Authority to assess the quality of the waterways in Mombasa Harbor concluded that "the ecology of the marine waters seems generally to be completely determined by natural processes." But the impact of human activities may gravely interfere with these processes.

Developments in Mombasa Harbor that will have an effect on the marine environment include an increase in shipping activities, dredging of the harbor and installation of industries producing poisonous effluents. The main threat to the environment from shipping is the disposal of ship wastes and the limited amount of control that can be exercised by the Mombasa Port Authority. Of particular concern is the disposal of oily waste, mainly from tankers, but also from other ships carrying fuel oil bunkers.

The Port Authority development plan up to 1988 recommended some dredging works in Mombasa Harbor to improve its entrance. This required blasting, which could have a permanent effect on the adjacent coral reefs, particularly if the dislodged material is dumped anywhere inside the harbor, rather than in deeper water outside the entrance. This could compound the problem of silting, which is already a major contributing factor to environmental degradation.

Along with the harbor expansion goes the intensification of industrial development. A pollution study of Mombasa Harbor showed that, until now, there are no major effluent contamination problems. But strict planning and a system of control need to be adopted and rigid standards applied if serious problems in the future are to be avoided.

A proposal for the construction of an offshore single buoy mooring (SBM) terminal for crude oil imports off the Mombasa mainland, for example, requires very careful study. First, the state of SBM technology in guarding against oil spillage is far from adequate. Secondly, having an oil terminal in clear view of the tourist hotels, reminding them of the ever-present risk of pollution of the fine sandy beaches, will be strongly opposed by the tourist industry, the second largest industry in Kenya (after agriculture).

In view of these likely conflicts of interests, more comprehensive investigations should be carried out as part of the development planning of Mombasa Harbor. KMFRI finds itself in a unique position in this regard and has taken steps to initiate the required studies.

Looking Ahead

With the establishment of an Exclusive Economic Zone, Kenya plans to develop sufficient capabilities not only to cope effectively with applications of foreign countries to conduct marine scientific research in its national waters, but also to formulate and implement an effective marine science policy of its own. KMFRI is well placed to play a significant role with regard to marine scientific research and the protection and pres-

Oyster culture has been
developed as part of a
biological oceanography
program undertaken by
Kenyan and Belgian scientists.
Photo: KFMRI

ervation of our marine environment. In an effort to meet these challenges, KMFRI has recently introduced both physical and geological oceanography to supplement the existing fisheries and environmental studies.

The Kenyan coast is subject to moderate wave energy under the influence of the "east coast swell," and to more severe waves generated by tropical cyclones. The large variability in shelf width influences local wave and climate processes. One of the unique aspects of the Kenyan coast is the high tidal range, approaching four meters. Thus, the highly dynamic conditions and large coastal variability require new approaches for the understanding and the management of the abundant coastal resource. The continental shelf and the shore zone are floored by mud, sand, and rock. Charts of the distribution of these materials will considerably aid the efficiency of bottom trawling for fish, mollusks, and crustaceans. Similarly, the samples of rock will permit an extension of geological maps of the land to the adjacent continental shelf. These maps will aid in the exploration for oil, gas, phosphate, and other economically interesting minerals.

Most information about the structure and stratigraphy of the continental shelf has been obtained by the oil companies, which tend to keep this information secret. Similar but less detailed information of the same sort will be obtained by our oceanographic re-search vessels and will be useful for inferring the origin and history of continents and ocean basins. Much has yet to be learned about the source beds, reservoirs, structural traps, and thickness of strata beneath the continental shelf of Kenya.

In the open ocean beyond the continental shelf, KMFRI will sample mainly suspended matter in order to study their concentrations and distribution. Such information will enhance the search for deep-sea fishery resources, both in terms of areas of concentration of fish and the seasonal variations that depend on the monsoons. Suspended matter consists of both organic and inorganic components. The organic components are especially abundant in areas and periods of upwelling; the inorganic are most concentrated after periods of maximum stream runoffs (floods). KMFRI will measure seasonal distributions and concentrations of both kinds of suspended matter as an aid to commercial fisheries.

With the growing awareness that the oceans are the common heritage of mankind, it is our hope that a better understanding of the nature of the oceans and their resources will be achieved through national efforts and international cooperation in marine research and development.

In recent years, the countries of the region have made remarkable progress in oceanography. The establishment of KMFRI is a prime example of this.

Notes on Contributors

Agustin Ayala-Castañares was born in Mazatlán, Mexico, in 1925. He obtained his Ph.D. at the National University of Mexico in 1963. He is considered one of the fathers of oceanography in Mexico and has devoted his life to the development of oceanographic capability in his country. He was Coordinator of Scientific Research at the National Autonomous University of Mexico from 1973 to 1980, Director of the Institute for Marine Sciences and Limnology at that University from 1981 to 1989, National Director for the National Plan to Create an Infrastructure for Marine Science and Technology in Mexico, a project carried out in cooperation with UNESCO and the United Nations Development Program (UNDP). He was also the first Vice-Chairman of the Intergovernmental Oceanographic Commission of UNESCO from 1975 to 1977, and its Chairman from 1977 to 1982. He was the first President of the College of Biologists of Mexico, as well as President of the Mexican Society of Natural History. Since 1968 he has been a Research Associate of the Scripps Oceanographic Institution. Dr. Ayala is the author of over forty published scientific papers in the fields of micropaleontology, marine ecology and geology, with particular interests in coastal lagoons in nearshore areas, their evolution and ecology, as well as the paleontological interpretation of ancient basins. In 1980, Dr. Ayala was decorated as an Officer of the Order of the Academic Palms of the French Ministry of the Universities.

Jacqueline Carpine-Lancre is Curator of the Library of the Oceanographic Museum of Monaco. She specializes in the bibliography and history of oceanography, with special emphasis on Albert I of Monaco.

Arthur G. Gaines, Jr. was born in New York. Dr. Gaines majored in biology and geology at Cornell University, and received his doctorate in oceanography from the University of Rhode Island in 1975. He taught chemistry and marine science at Atlantic College in Wales (1975–77) and was Staff Scientist at the Sea Education Association in Woods Hole from 1977 until 1979, during which time he was Chief Scientist on five cruises of the R/V *Westward*. Subsequently, he joined the staff of the Woods Hole Oceanographic Institution, where he is

presently a staff member at the Marine Policy Center. His special interest is in the societal value of marine science, and he conducts research on improved coastal management and decision-making through the incorporation of scientific information and technology.

Gotthilf Hempel was born in Göttingen, Germany, in 1929. He is Professor of Marine Biology at the University of Kiel. He has also been instrumental in the establishment of a Center for Tropical Marine Ecology at the University of Bremen. He is the founder of the Alfred-Wegener institute for Polar and Marine Research which, in the 1980s, became the center of polar research in Germany and one of the world's leading institutes of this kind. Dr. Hempel attended the Universities of Mainz and Heidelberg and worked for several years at various marine laboratories in Germany, Norway, the United Kingdom and the United States. He conducts one or two research cruises every year, to study the distribution and ecology of fish and fish larvae, and the biology of krill. For more than thirty years he has been engaged in various nongovernmental and governmental organizations for marine and polar research. His special interest lies in the partnership between scientists of industrialized and developing countries.

Geoffrey L. Kesteven, an Australian, earned his first degree at Sydney University with a thesis on the comparative anatomy of lizards. His first appointment was as Research Officer for the New South Wales Department of Fisheries. He has devoted his life to fisheries research, most of it as a Scientific Advisor to the Food and Agriculture Organization of the United Nations (FAO) in Rome and, for many years, in Peru and Mexico. He is the author of numerous scientific papers, and a member of the Planning Council of the International Ocean Institute. Since his retirement from FAO, he has been the Director of a consulting firm on fisheries management in Australia.

Bosko D. Loncarevic is a Research Scientist with the Geological Survey of Canada at the Bedford Institute of Oceanography. He has been engaged in geophysical research and surveys for over thirty years and is the author of over fifty scientific publications. Dr. Loncarevic graduated from the University of Toronto and the University of Cambridge, specializing in the field of Marine Gravity measurements. He has worked for Rio Tinto of Canada and, since 1963, has been a staff member of the Bedford Institute of Oceanography, where he established a Marine Geophysics program with emphasis on multi-parameter surveys of the continental shelf off Canada, as well as research on the continental margin and in the deep ocean on the Mid-Atlantic Ridge. He has also lectured at Dalhousie University and at the University of California at San Diego. Between 1969 and 1972 he was Assistant Director for Research at BIO and, from 1972 to 1977, he was the first Director of the Atlantic Geoscience Center at BIO. He has been involved in international joint projects, expeditions, and working groups. In 1977 he was the Resident Director of the Earth Physics School at the International Center for Theoretical Physics in Trieste, Italy, and he was a member of the Scientific Advisory Committee of the International Geological Correlation Program associated with UNESCO in Paris for six years. He is a Director of the Huntsman Foundation.

Eric L. Mills is Professor of Oceanography and Biology at Dalhousie University, Canada, where he works on the history of the marine sciences, especially of biological and physical oceanography.

Andrei S. Monin, born in Moscow in 1921, obtained his Ph.D. and his D. Sci. at the University of Moscow. He specialized in geophysical fluid dynamics, including hydrodynamics of the atmosphere and ocean and planetary interiors. He has worked at the Academy of Sciences of the U.S.S.R. since 1951, and was Director of the P. P. Shirshov Institute of Oceanology from 1965 to 1987. He is the author and co-author of twenty monographs in Physical Oceanography, including *Variability of the Ocean* (1974), *Turbulence in the Ocean* (1981), and *Introduction to the Theory of Climate* (1982). Since 1972 he has been a correspondent Member of the Academy of Sciences of the U.S.S.R. and, since 1973, a Foreign Honorary Member of the American Academy of Science and Arts. He is also a Foreign Associate of the National Academy of Science of the United States and, in 1986, he was awarded a Ph.D. *Honoris Causa* by the University of Göteborg, Sweden.

Takahisa Nemoto was born in Japan in 1930 and died August 22, 1990. He graduated from the Faculty of Agriculture of the University of Tokyo in 1953, specializing in biological oceanography. His research in

this area was conducted at the Ocean Research Institute of Tokyo University, where he was Director General from 1986 until the time of his death. Professor Nemoto served as special advisor to the Japanese Government as a member of the Council for Science, the Council for Ocean Development, the Japanese Council for Science and Technology, and the Council for Geodetics. (For the general information contained in the opening pages of this chapter, I am indebted to my student, Hirose Hirotaki. E.M.B.)

Eric Onyango Odada was born in Kenya in 1944. He obtained his Master's degree in Geological and Mineralogical Sciences at the Leningrad Mining Institute in the U.S.S.R. in 1970. During the ensuing years he continued his studies, taking courses in Mineral Exploration and Mining in Developing Countries, at the Loeben Mining Institute in Austria, and in Management and Conservation of Marine Resources at the International Ocean Institute of Malta. He was awarded his Ph.D. at the Imperial College for Science and Technology of London University and his D.I.C. in Applied Marine Geochemistry related to Mineral Exploration at Sea. Dr. Odada has been Exploration Geologist for the Department of Mines and Geology in Kenya and a Senior Geologist in charge of the Coastal Province of Mombasa, Kenya. He has been Chief Geologist responsible for Mineral Resources Exploration and Development at the Ministry of Environment and Natural Resources, and Principal Research Officer in charge of nonliving aquatic resources at the Kenya Marine and Fisheries Research Institute, Mombasa. He is now a Professor at the University of Nairobi. He was awarded a UNESCO Marine Science Fellowship and a British Foreign and Commonwealth Scholarship for research at the Imperial College in London.

Qin Yunshan is a Senior Scientist at the Institute of Oceanology of the Chinese Academy of Sciences. He was born in Shandong Province in 1933 and graduated from the Beijing College of Geology. He is Director of the Chinese Society of Oceanology and Limnology and of the Chinese Society of Oceanography. He is also Vice President of Working Group-II (East and Southeast Asia) of the Pacific and Indian Ocean Subcommission of the Commission on Quaternary Shorelines. Dr. Qin is the Editor of *Geology of the Bohai Sea*, (Beijing: Science Press, 1989), and of *Geology of the East China Sea*, (Beijing: Science Press, 1988). He has published over forty papers in Chinese and international journals. Some of the subjects covered are: "Primary study on the relief and sediment type of the continental shelf of China," (*Oceanologia et Limnologia sinica*, 1963); "Buried paleoriver systems in the western part of the South Huanghai Sea," (Kexue Tongbao, 1986); "Study on suspended matter in seawater in the southern Yellow Sea," (Chinese Journal of Oceanology and Limnology, 1988, in English). His major research interests are in sedimentation, sediment type, distribution pattern, and sediment composition of the China continental shelf. From these investigations, the sedimentation model of the China shelf seas and a sediment distribution map have been compiled. His other interests include suspended matter in seawater, geo-hazard phenomena in the seabed, and marine quaternary geology. The Chinese Academy of Science awarded him a second prize for his work on geology of the Bohai Sea (1986) and a first prize for his study on the Geology of the East China Sea (1988). The National Economic Commission awarded him the Nation's First Prize for his Regional Geotechnic Survey and Evaluation of the Oil Exploration Area in the Western South China Sea (1988).

Tumkur S. Shaila Rao was born in Chikkaballapur, India, in 1926. He obtained his doctorate of Science in Marine Ecology and Biological Oceanography at Andhra University, India. In 1956 he went to the United States for training in oceanography at the Woods Hole Oceanography Institution under a UNESCO fellowship. In 1962 he was invited to work as Scientific Liaison Officer for the U.S. Program in Biology during the International Indian Ocean Expedition and, in this capacity, he managed the cruise planning of the R/V *Anton Brunn* in the Indian Ocean. He was instrumental in organizing the Regional Center of a new Oceanographic Institute at Bombay. From there, he moved on to Cochin as head and scientist-in-charge of the Indian Ocean Biological Center established in collaboration with UNESCO to process and analyze the International Indian Ocean Expedition's collection of zooplankton. He supervised this work and prepared and published a set of plankton atlases for the Indian Ocean, the first of this kind in the world. In 1978 he became Assistant Director, and in 1981, Deputy Director, of the National Institute of Oceanography in Goa. As Chairman of the Ship Committee, he managed the research vessels *Gaveshani* and *Saga Kanya*. With *Saga Kanya* he traveled in 1983 as Chief Scientist on her maiden voyage to India from West Germany. Dr. Rao is now the Head of the Department of Marine Science and Marine Biotechnology, and the Dean of the Faculty of Applied Sciences at Goa University.

Roger Revelle was born in Seattle, Washington, in 1909 and he died July 15, 1991. He graduated from Pomona College in 1929 and received his Ph.D. in Oceanography from the Scripps Institution of Oceanography of the University of California in 1936. After receiving his degree, he was appointed an instructor on the faculty of Scripps. He became Professor of Oceanography at that Institution in 1948, and its Director in 1951. During the following ten years, Scripps experienced very rapid growth in the size and breadth of its scientific research and teaching, becoming the largest oceanographic institution in the world. During the latter part of his tenure as Director, Roger Revelle was much involved in the establishment of the University of California Campus in La Jolla. The University Regents named the first college of this new campus after Revelle. He resigned the directorship of Scripps in 1964 to become Professor of Population Policy and Director of the Harvard Center for Population Studies at Harvard University. In 1978 he returned to the University of California, San Diego, as Professor Emeritus of Science and Public Policy, a position in which he was still active at the age of 81. Roger Revelle was one of the founders of the Intergovernmental Oceanographic Commission of UNESCO, the Scientific Committee on Oceanic Research of ICSU, and the International Foundation for Science. From 1972 to 1990, he was a Member of the Board of Trustees of the International Ocean Institute.

John H. Vandermeulen is a Research Scientist with the Department of Fisheries and Oceans at the Bedford Institute of Oceanography. His research interests include the biochemistry and physiology of petroleum hydrocarbon stress, the chemistry of oil weathering, the fate and persistence of pollutants in coastal environments, and environmental pollutant monitoring. Dr. Vandermeulen is also Adjunct Professor at Dalhousie University (Halifax, Nova Scotia) in the School of Resource and Environmental Studies, where he teaches a regular graduate course, "Marine Resource Management in the Coastal Zone." He is an Associate of the International Ocean Institute (Malta), Research Associate of the Oceans Institute of Canada, and a Director of the International Center for Ocean Development. Dr. Vandermeulen has co-authored three books on aquatic pollution and is co-editor of *Ocean Technology, Development, Training and Transfer* (Pergamon Press), a study of environmentally sustainable and socially relevant technologies in the marine sector through new forms of private and public international cooperation.

Selected Bibliography

Academy of Sciences of the U.S.S.R. *Academy News* (Vestnic Akademii Nauk), journal. Moscow: U.S.S.R., Academy of Sciences, 1931.

Academy of Sciences of the U.S.S.R. *Atmosphere and Ocean Physics* (Fizika atmosfery i okeana), monthly journal. Moscow: U.S.S.R. Academy of Sciences, 1964.

Academy of Sciences of the U.S.S.R. *Marine Biology* (Biologiya morya), bimonthly journal. Moscow: U.S.S.R. Academy of Sciences, 1975.

Academy of Sciences of the U.S.S.R. *Journal of the U.S.S.R. Academy of Sciences* (Doklady Akademii Nauk SSSR), journal. Moscow: U.S.S.R. Academy of Sciences, 1933.

Academy of Sciences of the U.S.S.R. *Oceanology* (Oceanologiya), journal. Moscow: U.S.S.R. Academy of Sciences, 1961.

Albert 1er, Prince de Monaco. "Voyages scientifiques du yacht *Princesse Alice*." *Report of the Sixth International Geographic Congress, held in London, 1895* (1896): pp. 437–41.

———. *La carrière d'un navigateur.* Monaco: Editions des Archives du Palais Princier, 1966, pp. xxii–239.

———. "Recueil des travaux publiés sur ses campagnes scientifiques par le Prince Albert 1er de Monaco." *Résultats des campagnes scientifiques accomplis sur son yacht par Albert I Prince Souverain de Monaco* (1932): p. 369.

———. "Séance solenelle d'inauguration, 29 mars 1910. Discours de S.A.S. le Prince de Monaco." *Discours prononcés à l'occasion des fêtes d'inauguration du Musée Océanographique de Monaco, 29 mars 1910, 1er avril 1910,* pp. 5–11.

———. "Le progrès de l'océanographie." *Revue scientifique* 5, 1 (6), (1904): pp. 161–166.

Allen, W. E. "The Growth of a Marine Observatory." *Internationale Review der Gesamten Hydrobiologie und Hydrographie* 39 (1939): pp. 464–71.

American Geophysical Union. *Oceanology of the Academy of Sciences of the U.S.S.R.,* Bimonthly, trans. 1965. Washington, D.C.: American Geophysical Union, 1965.

American Oceanographic Delegation. *Oceanography in China: A Trip Report of the American Oceanographic Delegation,* submitted to the Council on Scholarly Communication with the People's Republic of China.

Washington, D.C.: National Academy of Sciences, 1980, p. 106, illustrated.

Anikouchine, W. A., and R. W. Sternburg. *The World Ocean: An Introduction to Oceanography.* Englewood Cliffs, N.J.: Prentice-Hall, 1973, pp. xi–338, illustrated.

Ayala-Castañares, A. "Las Ciencias del Mar y el Desarrollo de Mexico." *Ciencia y Desarrollo* 7 (1973): pp. 6–28, illustrated.

———. "The Role of Universities in Building National Capability in the Marine Sciences." *Impact of Science on Society.* Nos. 3, 4. Paris: UNESCO (1984): pp. 405–410, illustrated.

———. "The Role of Universities in Building National Capability in the Marine Sciences." Chap. 22 in *Managing the Ocean: Resources, Research, Law,* edited by J. C. Richardson. Paris: UNESCO, 1985, pp. 223–230, illustrated.

———. Informe de Labores, 1981–1987. Mexico City: Inst. Cienc. Mar y Limnol., 1987.

———, E. Mendez Palma, R. Mendoza de Flores, D. A. Ortega, and D. Sepulveda. "Estructura y Evolucion de la Investigacion Cientifica en la UNAM, (1929–79)." In *La Investigacion Cientifica de la UNAM 1929–1979.* Serie conmemorativa del Cincuentenario Autonomia. Mexico City: Univ. Nal. Aut. Mexico, 1979, pp. 13–18.

———, R. Mendoza de Flores, J. A. Nieto, and D. Sepulveda, "Estructura y Evolucion de la Investigacion Cientifica en la UNAM." *Ciencia y Desarrollo* 34 (1980): p. 33–48.

Bascom, Willard, N. A. *A Hole in the Bottom of the Sea.* New York: Doubleday, 1961, p. 352.

Bedford Institute of Oceanography. *BIO Review.* Dartmouth: Bedford Institute of Oceanography, 1984, 1985, 1986.

———. *Science Review.* Dartmouth: Bedford Institute of Oceanography, 1987.

Behrman, Daniel. *Assault on the Largest Unknown.* Paris: UNESCO Press, 1981, p. 96, illustrated.

Berlin, G. et al. *East African Port Development Study.* Kenya Port Authority Report, vol. 4, 1977, p. 116. Mombasa: unpublished.

Bonnefous, E., and P. Roy. *Institut océanographique, Fondation Albert I Prince de Monaco.* Paris: Institut

Océanographique; Monaco: Musée Océanographique, 1981, p. 55.

Britt, Albert. *Ellen Browning Scripps: Journalist and Idealist.* Oxford: Printed for Scripps College at the University Press, 1960, p. 134.

Brown, R. G. *Voyage of an Iceberg.* Toronto: Lorimer, 1986, p. 152.

Brush, Steven G., and Helmut E. Landsberg, with the assistance of Martin Collins. *The History of Geophysics and Meteorology: An Annotated Bibliography.* New York & London: Garland Publishing, 1985, p. 450.

Bullard, Edward C. "The Emergence of Plate Tectonics: A Personal View." *Annual Review of Earth and Planetary Science* 3 (1975): pp. 1–30.

Buzzati–Traverso, A. A. *Perspectives in Marine Biology.* Berkeley & Los Angeles: University of California Press, 1958, p. 621.

Carpine, C. "Catalogue des appareils d'océanographie en collection au Musée Océanographique de Monaco. 1. Photomètres. 2. Mesureurs de courant." *Bulletin de l'Institut Océanographique, Monaco* 73 (1437) (1987): p. 144.

Chang, Wen-yu. *The Marine and Continental Tectonic Map of China and Its Environs.* Beijing: Science Press, 1983.

Chinese Association of Oceanography. *Collection of Essays on the Marine Development Strategy in China* (in Chinese). Internal use only. Beijing, 1985.

Chinese Journal of Oceanology and Limnology, journal (in English). Beijing: Ocean Press, 1971.

Consejo Nacional de Ciencia y Tecnologia. "Programa nacional Indicativo para el Approvechamiento de los Recursos Marinos." *Ciencia y Tecnologia para el Approvechamiento de los Recursos Marinos (Situacion actual, problematica y politicas indicativas).* Mexico City: CONACyT, 1982, p. 115, illustrated.

Cox, Donald W. *Explorers of the Deep: Pioneer Oceanography.* Maplewood, N.J.: Hammond, 1968, p. 93.

Department of Fisheries and Oceans. *Canada's Oceans: An Economic Overview and a Guide to Federal Government Activities.* Published by the Information and Publications Branch, Department of Fisheries & Oceans, Ottawa, Ontario, 1986, p. 56.

Dinsmore, Robertson. "The University Fleet." *Oceanus* 25 (Spring 1982). Woods Hole, Mass.: Woods Hole Oceanographic Institution, 1982, pp. 5–14.

Edmonds, A. A. *A Voyage to the Edge of the World.* Toronto: McLelland & Stewart, 1973, p. 254.

Ewing, Gifford C., ed. *Oceanography from Space: Proceedings of Conference on the Feasibility of Conducting Oceanographic Explorations from Aircraft, Manned Orbital and Lunar Laboratories. Held at Woods Hole, Mass., August 24–28, 1964.* Woods Hole, Mass.: Woods Hole Oceanographic In-

stitution, ref. no. 65–10, April 1965, p. 469, illustrated, index.

Fairbridge, R. W. *Encyclopaedia of Oceanography.* New York: Reinhold Pub. Co., 1966, pp. xiii–1021, illustrated, maps.

Filmore, S., and R. W. Sandilands, *The Chartmakers: The History of Nautical Surveying in Canada.* Toronto: NC Press Ltd., 1983, p. 256.

Fleischer, A. A., and J. Contreras Urruchua. "Censos de ballenas grises (*Eschrichtium robustus*) en Baha Magdalena, B.C.S." Mexico: *Ciencia Pesquera. Inst. Nal. Pesca. Serie pesca. Mexico* 5 (1) (1986): pp. 51–62, illustrated.

Foster, M., and C. Marino. *The Polar Shelf.* Toronto: NC Press Ltd., 1986, p. 128.

Golantry, Eric. "A Profile: The Scripps Institution of Oceanography." *Oceans* 10 (November/December 1977): pp. 2–7.

Gorshkov, S., ed. *Atlas of the Oceans.* 3 vols. Moscow: Academy of Sciences of the U.S.S.R., 1975–79.

Government of Kenya. *Science and Technology Act, Laws of Kenya.* Chap. 250 in rev. ed. Nairobi: Government Printer, 1980.

Hatchey, H. B. *Oceanography and Canadian Waters.* Ottawa: Fisheries Research Board of Canada, 1961, p. 120.

Helms, Phyllis B., "Oceanographic Expeditions: Names and Notes." SIO Ref. no. 77–13 (July 1977), p. 21.

Hsu, Kenneth J. *The Great Dying.* San Diego: Harcourt Brace Jovanovich, 1986, p. 292.

Hutchinson, G. Evelyn. "Guano Deposits." *Bull. Amer. Mus. Nat. Hist.* 96 (1950): pp. 1–554.

Johnstone, K. *The Aquatic Explorers: A History of the Fisheries Research Board of Canada.* Toronto: University of Toronto Press, 1977, p. 342.

Kamenkovich, V., M. Koshlyakov, and A. Monin. *Synoptic Vortices in the Ocean.* Leningrad: Hydrometeoizdat, 1982, p. 264. 2d ed., 1987, p. 510. (Eng. ed., London: Reidel, 1986, p. 433.)

Keeling, Charles D. "Is Carbon Dioxide from Fossil Fuel Changing Man's Environment?" *Proceedings of the American Philosophical Society* 113 (February 1970): pp. 10–17.

Kenya Marine and Freshwater Research Institute (KMFRI). *Aquatic Resources of Kenya.* Proceedings of the Workshop of the KMFRI. Mombasa, Kenya, 1981.

Kenya Marine and Freshwater Research Institute (KMFRI). *Kenya Aquatica,* KMFRI Quarterly Bulletin. Nos. 1–3. (1985). Kenya: Mombasa, 1985, p. 115.

Kofoid, C. A. "The biological stations of Europe." *Bulletin, United States Bureau of Education* 4 (1910): pp. xiii–360.

LaGuarda–Figueras, A., and J. L. Rojas Galaviz. "Centro de Ciencias del Mar y Limnologia." In *La Investigacion Cientifica de la UNAM, 1929–1979*, pp. 335–385. Serie Conmemorativa Cincuentenario Autonomia, vol. 5. Mexico City: Univ. Nal. Auton., Mexico, 1987.

Lasky, Marvin. "Historical Review of Underwater Acoustic Technology: 1939–1945 with Emphasis on Undersea Warfare." U.S. *Navy Journal of Undersea Acoustics* 25 (October 1975): pp. 885–918.

Lillie, F. R. *The Woods Hole Marine Biological Laboratory*. Chicago: University of Chicago Press, 1944, p. 284.

Luo, Yu-ru, and Sheng-kui Zheng, eds. *Marine Undertakings of Modern China* (in Chinese). Beijing: China Social Science Press, 1985.

Macnae, W. "A General Account of the Fauna and Flora of Mangrove Swamps and Forests in the Indo-West-Pacific Region." In *Advances in Marine Biology*, vol. 6, pp. 73–270. London and New York: Academic Press, 1968.

————. "Mangrove Forests and Fisheries." Indian Ocean Fishery Commission, Indian Ocean Programme, IOFC/DEV/74/34, Rome: FAO, 1974, p. 35.

Maienschein, Jane. "History of Biology." In *Historical Writings on American Science*, OSIRIS, edited by Sally Gregory Kohlstedt and Margaret W. Rossiter, 1985, pp. 147–162.

Marine Science, journal. Beijing, People's Republic of China.

Mazur, Allan, and Elma Boyko. "Large-Scale Ocean Research Projects: What Makes Them Succeed or Fail?" *Social Studies of Science*. London & Beverly Hills, Calif.: SAGE Publications, 1981, pp. 425–449.

McEvoy, Arthur F. *The Fisherman's Problems: Ecology and Law in the California Fisheries, 1850–1980*. New York: Cambridge University Press, 1986, p. 368, bibliography.

Menard, Henry William. *The Ocean of Truth: A Personal History of Global Tectonics*. Princeton: Princeton University Press, 1986, index, bibliography.

Monin, A., and R. Ozmidov. *Turbulence in the Ocean*. Moscow: Academy of Sciences of the U.S.S.R., 1981.

Monin, A., V. Kamenkovich, and V. Kort. *Changes in the World Ocean*. Moscow: Academy of Sciences of the U.S.S.R., 1974.

Muller, D. G. *China As a Maritime Power*. Boulder, Colorado: Westview Press, 1983.

National Council for Science and Technology (NCST). *Science and Technology for Development*. Rep. no. 4, 1980. Nairobi, Kenya: 1980, p. 175. Unpublished.

Nierenberg, William, "Oceanography." In *The University and Applied Research: A Symposium on the Occasion of the Fortieth Anniversary of the Applied Research Laboratory, The Pennsylvania State University, October 1, 1985*, pp. 56–61. Pennsylvania State University, 1985.

Norconsult SA for Mombasa Municipal Council. *Mombasa Water Pollution and Disposal Study*. Prelim. rep. Mombasa, 1975.

Oceanologia et Limnologia Sinica, journal. Beijing: People's Republic of China.

Oppenheimer, J. M., "Some historical backgrounds for the establishment of the Stazione zoologica at Naples." In *Oceanography: the Past*, edited and translated by M. Sears and D. Merriman, pp. 179–187. New York: Springer Verlag, 1980.

Ostrom, B. "Marine Research in China." 267 *Nature* 30 (June 1977): pp. 794–97.

Otto, L. A. "China and the Law of the Sea." Ph.D. thesis, University of British Columbia, 1981.

Oxner, M. "Le nouvel Aquarium du Musée de Monaco." *Sud, magazin méditerranéen* 5 (74): 1932, pp. 26–29.

Panikkar, N. K., and T. M. Srinivasan. "The Concept of Tides in Ancient India." *Ind. Jour. of Hist. Sci.* Vol. 5, no. 1 (1971): pp. 36–50.

Pounder, E. R. *Physics of Ice*. Oxford: Pergamon Press, 1965, p. 151.

Prescott, William Hickling. *History of the Conquest of Peru*. New York: The Modern Library, 1936, pp. xxxv–1288.

Raitt, Helen, and Beatrice Moulton. *Scripps Institution of Oceanography: First Fifty Years*. San Diego: Ward Ritchie Press, 1967, p. 217.

Revelle, Roger, "How I Became an Oceanographer and Other Sea Stories." *Annual Review of Earth and Planetary Sciences* 15 (1987): pp. 1–23.

Richard, J. "Ma première campagne océanographique avec le Prince Albert." *Les Amis du Musée Océanographique de Monaco*. 1947 (2), pp. 1–11; (3), pp. 1–11; (4) pp. 1–15.

————. "L'Institut océanographique de Monaco et ses ressources de travail." En section des sciences, *Comptes rendus du 71e Congrès des sociétés savantes de Paris et des départements tenu à Nice en 1938*, pp. 297–300.

Rodriguez, Cynthia. "On Being a Woman at Sea." In *Bear Facts* 13 (December 1974): p. 2.

Salkovitz, Edward I., ed. *Science, Technology, and the Modern Navy, Thirtieth Anniversary, 1946–1976*. Arlington, Virginia: Department of the Navy, Office of Naval Research, 1976, p. 569.

Schlee, Susan. *On Almost Any Wind: The Saga of the Oceanographic Research Vessel Atlantis*. Ithaca, N.Y.: Cornell University Press, 1978, p. 301.

————. *The Edge of an Unfamiliar World: A History of Oceanography*. New York: E. P. Dutton, 1973, p. 398.

Schoff, W. H., *The Periplus of the Erythrean Sea: Travel and Trade on the Indian Ocean*. Translated from the Greek into English. New York: Longmans Green and Co., 1912, p. 323, illustrated.

Sears, M., and D. Merriman, eds. *Oceanography: The Past.* New York: Springer Verlag, 1980, p. 812.

Secretaria de Gobernacion. "Decreto por el que se adiciona el articulo 27 de la Constitucion Politica de los Estados Unidos Mexicanos, para establecer una zona economica exclusiva situada fuera del mar territorial." *Diario oficial.* (6 February, 1976): p. 2.

————. "Ley Federal del Mar." *Diario oficial.* (8 January 1986): pp. 3–9.

————. "Decreto por el que se aprueba la Convencion de las Naciones Unidas sobre el Derecho del Mar." *Diario oficial.* (18 February 1983): p. 2.

Shirshov Institute of Oceanology. *Oceanology.* 10 vols. Moscow, 1977–80.

————. *The Pacific.* 10 vols. Moscow: Izdadelstvo "Nauka," 1966–1974.

Smith, M. L., ed. *Woods Hole Reflections.* Woods Hole, Mass.: Woods Hole Historical Collection, 1983, p. 301.

Soviet Navy, Central Administration for Navigation and Oceanography. *Marine Atlas.* 2 vols. Moscow: Morskoi Atlas Publishers, 1950–53.

Studia Marina Sinica, journal. Beijing: People's Republic of China.

Vallaux, C. "Jules Richard, l'homme, le savant, l'organisateur." *Bulletin de l'Institut Océanographique, Monaco* 43 (892), (1946): pp. 11–15.

Walton, A. "Le Laboratoire international de radioactivité de Monaco: historique et activités." *Bulletin— Agence internationale de l'énergie atomique* 29 (3) (1987): pp. 57–59.

Warren, B. A. "Medieval Arab Reference to the Seasonally Reversing Currents of the North Indian Ocean." *Deep Sea Res.*, 13, 1966.

Watermann, B. *Bibliography of the History of Oceanography in Germany: Chronological List of Titles (1557–1986), and Index.* Hamburg: Deutsche Gesellschaft für Meeresforschung, 1987, pp. xiii–198, maps.

Williams, F. "Preliminary Survey of the Pelagic Fishes of East African Waters." Fisheries publications, no. 8. London, 1956, p. 68.

Yang, Jinsen, et al. *Brief History of Marine Fishery in China* (in Chinese). Beijing: China Ocean Press, 1982.

Yang, Jinsen, and Zhiguo Gao. *Marine Policy in the Asian-Pacific Region* (in Chinese). Beijing: China Ocean Press, 1990.

Zenkevich, Lev. *The Biology of the U.S.S.R. Seas,* translated by S. Botcharskya. Moscow: Akademii Nauk SSR, 1963, p. 739, maps. (London & Toronto: Allen and Unwin, 1963, p. 955, illustrated, maps.)

Index

A page number in *italics* refers to an illustration.

163, 169, 171
Syntex Corporation, 39
Sysoyev, N., 161
SZC Series STD Recorder, 203

Taft, Bruce, 43
Tahiti, 46
Tansei-Maru, 216; *220*
temperature, 76, 80, 144, 146, 188, 203
Terminos Lagoon, 253
Texas A & M University, 50
Third Institute of Oceanography, Xiamen, 199
Thomson, Charles Wyville, 122
Thoulet, Julien, 127, 129
Tonga (Kermadec) Trench, 25, 41–42, 50
Tong Dizhou (T. C. Tung), 199
Torrey Canyon, 98, 101
Trivandrum, India, 180
Tropical Ocean Global Atmosphere Program (TOGA), 237
Tsukamato, Dr., 222
tube worms, 161, 225; *157*; large vent (*Riftia pachyptila*), 34; *35, 36*
Tucholke, Brian, 86
tuna, 266, 270
Twin Otter aircraft, 114

UNCTAD, 12
UNDP, 266
UNEP, 266
UNESCO, 12, 177, 192, 247, 253, 266
unicorn fish, 260
United States Bureau of Fisheries, 65, 67
United States Coast and Geodetic Survey, 67
United States Fish and Wildlife Service, 45
United States Fisheries Commission, 57, 60–62
United States Geological Survey (U.S.G.S.), 57, 179
United States Light House Service, 57

University of California, 28, 31, 52
University of California Division of War Research (UCDWR), 22
University of Chicago, 62, 67
University of Hamburg, 143, 144
University of Tokyo, 216
University of Wisconsin, 67
"Upper Cave Man," 196
USAID, 266
U.S.S.R. Academy of Sciences, 158, 160
Utafiti, 265

Vacquier, Victor, 25
Valdivia, 138
Varuna, 179–80
Vedros, Neylan, 39
Vema, 108, 179
Vetter, Russell, 34
Victoria, Lake, 265, 270
Vine, Allyn, 63, 76, 89, 103; *79*
Vinogradov, M., 166
Vityaz, 161, 166, 169, 171, 178, 179
Von Herzen, Richard, 87

Wake Island, Northwestern Pacific, 25
Waksman, Selman, 67
Wang Chong (Lun Heng), 196
Ward Hunt Ice Shelf, 115
warm core ring, *222*
Watamu Marine Park, 261, 270
Watkins, Bill, 80
Wattenmeer (mud flats), 146
Weddell Sea, 144, 147, 149, 151
Wegener, Alfred, 153, 155; *154*
Western Indian Ocean, 50
Western North Pacific, 223
Western Pacific, 45, 46, 138, 224, 225
West Greenland, 151
Westward, 57
"Whale Island," Watamu Marine Park, 261
whales, 153; Gray, *238*
Whitman, C. O., 60, 62, 63
Wiebe, Peter, 80
Williams College, 62

Winant, Clinton, 46
Winnaretta-Singer, 132
Women's Education Association, 62, 63
Woods Hole Oceanographic Institution, Massachusetts, 22, 37, 42, 43, 56–94, 98, 177, 179, 246; *54, 55, 58, 59*; creation of, 63, 65; formation of the Fisheries Laboratory, 60–62; Gulf Stream research, 73; ocean temperature research, 76, 80; Project NOBSKA, 87, 92; Redfield's work, 73, 76; Rose-Lillie plan, 65, 67
Woods Hole Research Center, 57
Wooster, Warren, 28
World Meteorological Organization (WMO), 143
World Ocean, 161, 164, 167, 168, 173, 189
World Ocean Circulation Experiment, 56, 144
World Ocean Climate Exercise (WOCE), 98, 253
World War I, 129
World War II, 22, 76, 80, 103, 131, 160, 164
Worthington, Valentine, 73

Xiangyang 16, 208
Xiangyanghong 09, 208; *207*

Yayanos, Aristides, 33
Yellow Sea, 202
Yupanqui, Tupac, 230

Zenkevich, Lev (L. A.), 163, 164, 178
Zheng He, 196
Zhoukoudian, 196
Zilitinkevich, 168
ZoBell, Claude, 33
zooplankton, 114, 138, 144, 179, 235, 252, 253, 254
Zubov, Mr., 163
Zumberge, Mark, *30, 31*